Old Wives' Lore for Gardeners

Old Wives' Lore *for* Gardeners

Diana Ferguson

Michael O'Mara Books Limited

First published in Great Britain in 2021 by
Michael O'Mara Books Limited
9 Lion Yard
Tremadoc Road
London SW4 7NQ

A CIP catalogue record for this book is available from the British Library.

Papers used by Michael O'Mara Books Limited are natural, recyclable products made from wood grown in sustainable forests. The manufacturing processes conform to the environmental regulations of the country of origin.

ISBN: 978-1-78929-316-6 in hardback print format
ISBN: 978-1-78929-317-3 in ebook format

1 2 3 4 5 6 7 8 9 10

The remedies mentioned in this book are traditional folk ones and may not be suitable for everyone. Please seek medical advice before trying any of them if you are concerned.

Illustrations by Fay Miladowska
Designed and typeset by Claire Cater
Cover design by Jade Wheaton

Printed and bound by CPI Group (UK) Ltd, Croydon, CR0 4YY

www.mombooks.com

To my daughters, Blanche and Charlotte,
who have discovered the joys of gardening.

Contents

Introduction

If you want to be happy for a day, get drunk;
For a month, get married;
For life, be a gardener.

Chinese proverb

As we potter in our gardens, perhaps planting a few spring bulbs or growing some vegetables, we are continuing an important tradition that goes back not just to the time of our grandparents or even great-grandparents but to the beginnings of civilization itself. Let's take a journey back in time to prehistory. At that stage in evolution, human beings were nomadic hunter-gatherers: their food supply depended on what they could find, whether it be the wild animals they hunted for meat or the edible plants they were able to forage. That isn't to say that they had wholly unsophisticated palates and didn't try to improve the flavour of what they ate with a spot of seasoning: archeologists have found evidence that they chewed mustard seed with their meat. But they still had little control over their environment and were subject to the vagaries of nature.

Then, perhaps around ten thousand years ago, humankind experienced a lightbulb moment. According to the historian and mathematician Dr Jacob Bronowski, presenter and author of the classic television documentary series, *The Ascent of Man* (1973), this point occurred when people first discovered how to cultivate the wild grain that formerly they had only been able to gather. This was when farming and, by extension, gardening were born. The ability to grow our own food crops was as important a development perhaps as the domestication of the horse, which allowed humans to cover the ground at previously unknown speeds, or the invention of the wheel, which opened up totally life-changing methods of transport.

Think about it for a moment: if you can grow your own food, you can stay put. You no longer have to keep following wild herds or foraging for whatever edible plants you can find to supplement your diet. And if you can stay put, you have more time – time to settle down and build homes and establish villages, time to make beautiful objects that endure and to develop your culture, which you could not do before because you were always on the move and had to travel light. The villages you create subsequently turn into towns, which turn into cities, and what we know as civilization is born ('civil' and 'civilization' are related to *civitas*, the Latin word for 'city').

Methods of grain cultivation were applied to other kinds of plants, and the beginnings of gardening as we might recognize it today gradually came into being. The expertise of these early growers has been passed on down the centuries from one generation to the next. The old-timers who are the custodians of this ancient knowledge did not acquire it by means of 'book learning' but hands-on, through experience, and through what they learned from their own elders. In the same fashion, they have shared what they know: many a celebrity gardener has revealed how their love of gardening blossomed not at horticultural college but by a grandparent's side, as they learned the magic secret of bringing new life out of a tiny, hard black speck called a seed.

So who are these old-time gardeners, these repositories of time-honoured horticultural wisdom? They are the ordinary folk of long ago who grew their own vegetables, herbs and flowers in their small plots, and their descendants are with us today. You may even know one or two yourself – perhaps a grandparent, or the no-nonsense 'Old Girl' or 'Old Wife' who is out there in all weathers and seems to know more about plants than you imagine you ever will, or the quiet 'Old Boy' you've seen tending his allotment or greenhouse.

Canny, frugal, experienced, knowledgeable and full of common sense, these gardening experts have a down-to-earth approach to their craft. They avoid waste and have ingenious ways of repurposing existing objects. By the same token, they know there is no need to splash out on expensive gardening products or chemical treatments because Nature – and the kitchen cupboard – will provide all the help and raw materials they need. Ironically, their 'old ways' are now 'new ways', because they practised organic gardening before the term was even invented. Whether you have a large country garden, a small urban plot or perhaps no more than a balcony or window ledge, these age-old principles are still relevant today.

When you head out to do a spot of digging or planting or trimming back, you may not be aware that you are quietly feeding your soul at the same time. Gardening is not only a physical activity, but a spiritual experience too. Quietly working in our gardens or even tending the potted plants on our balcony can be almost a meditative experience, giving us a respite from the busyness and pressures of our everyday lives.

Research bears this out. People who spend time gardening or out in nature have been found to be in better health and have higher psychological well-being. Doctors are even starting to offer their patients 'green prescriptions', which recommend such activities as gardening, or visiting parks and green spaces, to treat anxiety, depression and loneliness.

So, next time you do a spot of gardening, take time to savour the experience, too. Sit for a while with a cup of coffee on a frosty morning, or listen to the birdsong on a warm spring day, and reconnect with the oldest healing source of all: Nature herself. Try it. It works.

1

The Gardener's Toolkit

There is a lovable quality about the actual tools. One feels so kindly to the thing that enables the hand to obey the brain. Moreover, one feels a good deal of respect for it; without it the brain and the hand would be helpless.

Gertrude Jekyll (1843–1932),

British horticulturalist, garden designer and writer

Gardening the old way is all about practicality and frugality, encapsulated in those two sayings, 'Waste not, want not' and 'Make do and mend'. The gardeners of old did not grow up in a throwaway society. If you can think of ways of upcycling (a very new but also very old concept), then do so. If you have to buy new, make sure that you care for the tools you have bought, and repair them where you can.

Basic toolkit

Only fools will lend their tools.

Traditional saying

Every gardener needs a few essential tools to make gardening easier and more pleasurable. Add to them as needed, and as finances allow. In the spirit of the old-fashioned philosophy of 'waste not, want not', you might be able to buy some tools second-hand, but do check that they are in good working order, and repair them if need be and you have the skills.

Hand fork – for weeding and lifting small plants; helpful for heavier soils

Trowel – narrow-bladed or wide-bladed; useful for digging small holes and planting out bedding plants and bulbs

Spade – for digging planting holes and lifting plants

Fork – used in a similar way to a spade, but easier than a spade on heavy soils

Hoe – brilliant for slicing through weeds

Secateurs – for pruning soft shoots and woody stems up to about 1 cm (½ in.) thick (always prune back to a joint and cut cleanly as ragged cuts could lead to infection)

Shears – for trimming hedges and cutting back perennial plants at the end of the growing season

Rake – for raking soil prior to sowing seed; a fan-shaped lawn rake is useful for collecting grass cuttings or raking up moss from lawns

Keep on hoeing

Regular use of a hoe will keep the weed population under control, and save you having to dig them up individually by hand. The trick with a hoe is to keep the blade flat, parallel with the soil surface; slide the blade across the top of the soil and into the weeds to sever the tops from the roots. With annual weeds, do this before they flower and start setting seed, or you'll be sowing the next generation of weeds. Regular hoeing can even be the death knell for persistent perennial weeds like bindweed. Deprived of their leaves – the little energy factories that convert sunlight into sugar – they will weaken. They may not totally die off, however, as even a small bit of root can regrow, so you'll need to keep up the treatment; eventually you will win.

> *When I go into the garden with a spade, and dig a bed, I feel such an exhilaration and health that I discover that I have been defrauding myself all this time in letting others do for me what I should have done with my own hands.*
>
> Ralph Waldo Emerson (1803–82), US author

Top-up toolkit

Add to your basic toolkit with these few extra items. There's nothing worse than having to scrabble around for what you need when you're in the middle of a job, so keep them within easy reach when you're gardening:

Gardening gloves – for protecting your hands while you work

Watering can – for a gentle sprinkle with the rose on, or a more directed jet of water with the rose off

Plant supports – bamboo canes or strong twigs for extra support to stop taller plants flopping over; tall canes and/or netting for climbers such as runner beans

String and scissors – for tying plants to plant supports

Wigwam
plant support
made of
bamboo

How to tie in a plant

When you secure a plant to a support with string, the string can sometimes dig into the stem as the stem pulls against it in the wind. To prevent this, wind the string in the shape of a figure 8:

1. Loop the string twice around the stem, winding it from the back to the front, then back again.

2. Tie a knot at the back of the loop, so you have a double loop of string wrapped snugly around the stem.
3. Tie the two loose ends of string to the support.

Now, the tension will be on the length of string between the loop and the support and not on the stem itself.

Supports for bushy plants

Swanky, ready-made 'cage' supports are available for bushy plants, such as peonies or sedums, to grow into. An alternative, old-fashioned solution is to place squares of wide-meshed rabbit wire over the shoots when they appear in spring (rather like putting chicken wire in a vase when you arrange flowers). The shoots will grow up through the netting, which may gently be raised as the plant grows. For extra support, slip a stake of some sort through the netting when the plant is about 23 cm (9 in.) high.

Old-timer's tip

When you first plant young trees, they will need tying in to stakes to support them until they are strong enough to stand alone. Branches heavily laden with fruit may also need tying in. Instead of buying special tree ties, save money by re-purposing old tights or socks.

Lawnmower or scythe?

A substitute for mowing with the scythe has lately been introduced in the form of a mowing machine… but it is proper to observe that many gardeners are prejudiced against it.

Jane Loudon, *The Ladies' Companion to the Flower Garden* (1841)

On 31 August 1830, an English mechanic called Edwin Beard Budding launched his new invention on an unsuspecting public. A cutting machine designed 'for the purpose of cropping or shearing the vegetable surface of lawns, grass-plats and pleasure grounds', it was in fact the world's first lawnmower. Budding got the idea for his contraption from observing the cross-cutting machines used in the textile mills where he worked. He is said to have tested his invention at night, for fear of the prying eyes and mocking comments of neighbours. Without his machine, would we have smooth-as-glass lawn tennis courts and bowling greens?

For centuries before the invention of the lawnmower, grass had to be cut by scythe, an implement consisting of a long-handled shaft called a 'snaith' and a long, curved, slicing blade. Scythes may have been in use as far back as 5000 BC, and are even depicted in Neolithic cave paintings found in Norway.

There were different types of scythe for specific tasks – cutting grass or wheat, clearing weeds or cutting reeds. Wielding a scythe was known as 'mowing', and it was a skilled task that took time to master. Moving together in harmonious rhythm, mowers worked their way across the land, clearing swathes of ground as they went – a 'swathe' being the name for the strip cut with one slice of a scythe. The following passage brings this picture of rural labour vividly to life:

> *He heard nothing but the swish of scythes, and saw before him Tit's upright figure mowing away, the crescent-shaped curve of the cut grass, the grass and flower heads slowly and rhythmically falling before the blade of his scythe, and ahead of him the end of the row, where would come the rest.*
>
> Leo Tolstoy, *Anna Karenina* (1878; trans. Nathan Haskell Dole, 1886)

The first wheelbarrow

The wheelbarrow is not thought to have made its appearance in Europe until the beginning of the thirteenth century. The Chinese were way ahead, however. Chuko (or Zhuge) Liang – an adviser to Liu Bei, founder of the Shu–Han Dynasty – is credited with inventing this useful piece of kit back in the third century, though illustrations dating from around AD 100 show a form of wheelbarrow in use well before he was born.

Tool care

'A bad workman blames his tools', as the old saying goes, so don't fall into this trap. Give your tools the respect they deserve; they are your allies in the garden, so care for them and they will serve you well.

Old-timer's tip

If you are thinking of throwing away a bristly
old doormat, don't! Use it as a tool scraper
instead: hang it up on your shed wall and scrape
tools against it to remove soil after use.

Cleaning

After using your tools, brush off any soil, leaves or grass clinging
to them, then wipe with an oily cloth to prevent rusting or
warping.

Storing

Don't rest hoes, spades or forks on their blades or prongs when
storing them, or they might blunt. Rest them on their handles
instead or, better still, hang them up on hooks on the shed wall.

Get innovative and improvise some storage solutions. An old
belt or strap, nailed at intervals along the wall, will create handy
loops for holding smaller tools; or you could store them in an old
wire vegetable rack, which will allow air to circulate to keep them
dry. Clean glass jars are ideal for holding really small items, such
as hooks, ties, labels, string – you'll be able to find them instantly
and see when you need to stock up again.

Sharpening

A good time to do this is before you put your tools away for the winter. You'll need a coarse brush, some medium sandpaper, oil (boiled linseed, for preference), a rag, and a metal rasp for filing.

1. Brush off any loose soil, then use sandpaper to remove any further dirt or rust.
2. Wipe over with oil.
3. Rub along both edges with the metal rasp. Oil the sharpened edges again.
4. Put your tools away in the shed or dry place. When you take them out again in spring, they should be rust-free and super-sharp – ready for action.

Old-timer's tip

Save money! Instead of buying a purpose-made kneeling pad, use an old, empty hot-water bottle for kneeling on when you are planting out or weeding.

Planters for free

This is something that would really appeal to frugal old gardeners: recycling so that nothing goes to waste. Almost any waterproof container can be pressed into service to hold plants – clean tins, old paint pots, that baking dish you no longer want, plastic ice cream tubs, that old bucket or wastebin... As long as the container is rigid enough, deep enough and has holes in the bottom for draining, it will do the job. For a rustic look, you can even use an old basket. Line it with plastic so that it holds the soil and is more waterproof, but remember to make drainage holes in the plastic.

You can pierce holes in foil containers. For ceramic containers, drill holes using the appropriate drill bit. Plastic might split if you pierce it, so heat a metal skewer over a flame and use it to melt the holes instead.

Whatever repurposed container you're using, remember to add a few broken bits of crockery or pebbles in the bottom to aid drainage.

Biodegradable seed pots

Save old cardboard egg boxes: the cups make perfect little pots for growing seedlings in. When the seedlings are ready to plant out, just pop them in the soil, cardboard cups and all. The cardboard will rot down in the soil, saving you the bother of having to turn the seedlings out, and protecting the roots from damage during transplanting. What could be more convenient?

Use cardboard toilet-roll tubes for growing sweet pea seedlings, which like a deep root run.

Like egg-box seed pots, plant them out tubes and all.

In the old days, milk and other liquids came in glass, not plastic, bottles, but there's no reason why you can't apply the waste-not-want-not approach to this more modern material. Cut large, clear plastic bottles in half and use the tops (with the necks uncovered, for ventilation) as cloches to protect young seedlings.

How to age plant pots

If you want a new terracotta pot to look like an antique urn, you can 'age' it artificially. Soak it in water, paint the outside with natural yogurt, leave it in a damp, shady place, and wait for moss and lichen to grow on the surface.

Old-timer's tip

Plants often die right down in the winter so you can't see where they are if you later want to move them, or add other plants around them. Before they totally disappear, a clever trick is to push a cane into the soil to mark the spot, perhaps tying a waterproof label or piece of coloured wool to the cane to identify the plant.

A gardening notebook

*I am trying to make a grey, green, and white garden. This
is an experiment which I ardently hope may be successful,
though I doubt it … All the same, I cannot help hoping that
the great ghostly barn owl will sweep silently across a pale
garden, next summer, in the twilight – the pale garden
that I am now planting, under the first flakes of snow.*

Vita Sackville-West (1892–1962), English author and creator of Sissinghurst
Castle Garden and its most celebrated feature, the White Garden

Keeping a gardening notebook is the ultimate gardener's tool.
This is the place to record and reflect on your successes, failures,
observations, aspirations and plans, rather like keeping a personal
journal. Many famous female gardeners have kept just such a
notebook. Gertrude Jekyll (pronounced 'Jee-kull'), who was one of
the most influential garden designers of the early twentieth century,
kept a series of scrapbooks, notebooks and photograph albums.

Another avid notebook-keeper was the aristocratic and wealthy
Vita Sackville-West, who discovered her passion for gardening after
she and her husband Harold Nicholson bought Sissinghurst Castle
in Kent, southeast England, in 1930. Over the ensuing thirty years,
she and a team of gardeners transformed what had been farmland
into one of the most renowned and respected gardens in the world,
including its most celebrated feature, the White Garden. Sackville-
West used her notebook as the basis for a series of articles she
produced for the *Observer* newspaper, from 1947.

Your garden is unlikely to be on such a grand scale as

Sissinghurst, nor as influential as Jekyll's garden at Munstead Wood, but you can still take great delight in keeping your own gardening journal. This is your notebook, so have fun with it and include whatever takes your fancy, such as:

- What worked and what didn't
- New plants you would like to add and changes you would like to make
- The likes and dislikes of specific plants
- Garden designs
- Sketches
- Photographs of favourite plants that you have grown
- Pictures from magazines that inspire you
- Seasonal notes

2

Down to Earth

The soil is the gift of God to the living.

Thomas Jefferson (1743–1826), third US President
and Founding Father of the United States

Forget the basalt columns of the Giant's Causeway in
Northern Ireland, the flowing rock layers of Antelope
Canyon in Arizona, the amazing sculpted dunes of the Sahara
… you have a geological mini-marvel in your own garden. It's
soil. Produced over eons by the Earth's natural processes, this
incredible substance is the basis of all life. It is the growing
medium for the plants that support the world's ecosystems and
sustain living creatures at all levels of the food chain, including
us humans. It deserves nurturing and care.

But you don't need to be a soil scientist to know how to look
after it. The trick, as always, is to work with Nature. With some
knowledge of the processes that produce good soil and some good

old-timer's know-how, you can have the kind of rich, crumbly loam that will keep your plants happy and thriving.

What is soil?

Put simply, soil is a mixture of minute particles of rock, organic matter, living organisms, air and water. Topsoil is the upper layer that is of most interest to gardeners, for it is there that our plants take root. Below this lies a denser and less fertile layer, called subsoil.

Depending on the mix of the different constituents, soil can be divided into six different types. In reality, plants are quite forgiving and will often grow in soils that are less than ideal for them, but it's useful to know your soil types if you want to improve them.

Clay soil Slimy and sticky when wet, this heavy soil drains poorly. Conversely, in hot weather it can dry rock-hard and crack like the surface of a desert. The good news is that it retains nutrients well, and roses like it.

Clues: if you roll this into a ball, it will hold its shape and feel like – well, clay!

→ Improve drainage and soil structure by digging in coarse sand and organic matter.

Chalky soil This is stony and drains well, but essential plant nutrients are easily washed out. Chalky soil is alkaline (the opposite of acid – think of limescale in your kettle). Plants that like free drainage and an alkaline soil, such as lavender and the cabbage family, will grow well in this.

Clues: try the 'Acid or alkaline?' soil test (on page 31).

🐛 Mulch to retain moisture and add fertilizer to top up nutrients.

Sandy soil This drains well and is easy to dig. On the down side, it dries out quickly and nutrients are easily washed out. Drought-tolerant plants such as sea holly, sedums and euphorbia, are suited to this kind of soil.

Clues: sandy soil feels gritty and, if rolled into a ball, crumbles easily.

🐛 Improve soil structure and nutrient content by adding organic matter such as manure or homemade compost.

Silty soil Made from fine particles, this drains well and retains more moisture and nutrients than sandy soil.

Clues: this feels smooth and rolls into a ball easily but does not hold its shape as well as a ball of clay soil.

🐛 Add organic matter to improve soil structure.

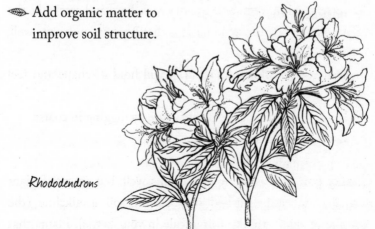

Rhododendrons

Peaty soil This acidic soil is dark in colour and low in nutrients, but contains plenty of organic matter. It's rare to find this in gardens – think more along the lines of peat bogs. Acid-loving plants such as rhododendrons and azaleas like peaty soil (which is why you should ideally grow such plants in acid, ericaceous compost). A note on peat: if you buy potting compost, go for a peat-free variety. Peat extraction damages important wildlife habitats and releases vast quantities of carbon dioxide into the atmosphere.

Clues: peaty soil feels spongy when you squeeze it.

Fertilize to top up nutrient levels.

Loamy soil If this your soil type, you've struck gold! Crumbly, free-draining and full of nutrients, this is the gardener's dream and what to aim for in improving other soil types.

Clues: this holds its shape when rolled into a ball, but not as well as clay soil.

Maintain this high-quality substance by topping up with organic matter.

Acid or alkaline?

If your plants aren't thriving, your soil may be too acid. Acid soil locks up nutrients so they are less available to plant roots. A time-honoured practice to correct this is to add garden lime (calcium carbonate), ground chalk or calcified seaweed to the soil, to make it more alkaline. An alkaline soil is particularly important for the cabbage family, because they don't like an acid environment. Adding lime also improves the structure of clay soil as it makes the tiny particles of clay clump together, allowing better drainage. Apply lime in winter, following the instructions on the bag for application rates and methods.

To get technical, a pH (the acid/alkaline reading) for soil of 7.0 is considered neutral. Above this, you have an alkaline soil; below, and your soil is acid. The ideal you are aiming for is a pH of 6.5, just on the acidic side. Nutrients are most freely available at this level, and it's also best for the activity of bacteria and earthworms.

Today, you can pop out to your local garden centre and buy an inexpensive soil-testing kit. In the old days, you might only have had a more homespun method like the following one. It won't be as accurate as a kit, but it will give you an idea.

1. Collect soil from different parts of your garden and put two spoonfuls in separate containers.
2. Add about 120 ml (4 fl oz) of white vinegar to one container. If the mixture froths up, your soil is alkaline.
3. If it doesn't froth, pour a little distilled water into the other container to cover the soil so it goes muddy, then add about 60 g (2 oz) of bicarbonate of soda. If the mixture bubbles and fizzes, your soil is acid.

Making compost

The time-honoured practice of making your own compost is a lesson in sustainability and exemplifies good, old-fashioned common sense. Instead of throwing away kitchen and garden waste and allowing it to go to landfill (where it would release methane gas as it was broken down), or even giving it to your local authority to turn into compost (involving transport by lorry), you can benefit by keeping your waste at home. Here – through the miracle of natural processes – it morphs into something completely different: a dark, rich, crumbly, sweet-smelling substance that is a real treat for your garden. Compost is a wonderful soil conditioner and great growing medium, and can also be used as a mulch. Perhaps best of all, though, it's totally free.

The compost bin

To start with, you'll need a container for your compost. You can, of course, buy one: there are plenty of good products on the market, including 'hot bins' which heat the contents to such a high temperature that they rot down much more quickly and, the manufacturers claim, turn into usable compost in just three months. They are not cheap, however.

If you want to be resourceful and save money, you can make your own bin from discarded wooden pallets. Clear a patch of earth, then set three pallets on it, fixed together at the sides, with a fourth pallet in front to act as a kind of gate so that you can access the compost whenever you want. This kind of bin takes up a fair amount of space, though, so you might prefer something smaller.

An old plastic rubbish bin with a lid can do the job, too. Drill holes around the bottom and outside of the bin, about 15–20 cm (6–8 in.) apart to help air flow and to give access to the little earth-workers, such as worms, that will help to break down the compost for you. Place your bin on bricks if siting it on solid ground to help aerate the contents; ideally, place it directly on the earth. If it's not too heavy to lift, the advantage of this type of container is that you can turn the compost by rolling the bin.

Compost heap

The raw materials

To kickstart the decomposing process, you need two key components in equal proportions to supply nitrogen and carbon. Fifty-fifty is the aim. Think half green (for nitrogen) – such as leafy material – and half brown (for carbon) – such as dry woody stems. Too much nitrogen-producing green stuff, especially grass clippings, and composting may happen too fast, resulting in a slimy green sludge; too little carbon, and composting may be too slow.

Do add:

- Leaves and old plants
- Grass clippings
- Annual weeds (as long as not in flower)
- Nettle leaves
- Vegetable and fruit peelings
- Tea bags
- Coffee grounds
- Crushed eggshells
- Straw
- Shredded paper (as long as it's not glossy)
- Shredded wool and cotton
- Dry stems
- Twigs and woody stems
- Cardboard, such as egg boxes

Don't add:

- Thick woody stems
- Autumn leaves (use to make leafmould instead – see page 36)
- Citrus fruits (these rot down slowly and are highly acidic, which reduces worm activity)
- Cooked food, bones, fat or eggs (they could attract rats)
- Raw meat
- Flowering annual weeds (their seeds could germinate in the compost)
- Tough perennial weeds such as bindweed (you'd need a higher heat to fully rot this down than you can achieve in a domestic compost bin)
- Plants infected with diseases, such as rust

- Manmade fibres
- Ash from coal fires
- Cat or dog faeces (as if you would!)

Getting started

1. Start by shredding the material. Break it up by hand, chop it, cut it – the smaller it is, the faster it will break down. You can scrunch up paper and leave egg boxes and toilet rolls whole, as these will create air pockets and help to aerate the mixture.

2. Layer the material into your bin and cover it to keep out the rain, then let Nature get started. Inside your compost, a vast population of invisible bacteria, fungi and other micro-organisms as well as worms and beetles will be working hard on your behalf, digesting the contents and generating heat as they do so. If you open up the top of the heap, you may even see steam drifting out.

3. To aid the process, your compost needs water and oxygen. Water lightly if necessary to keep it moist, and turn it – from the outside in – to aerate it. Do this every few weeks but don't add any fresh material after turning.

Warning: take great care when turning the mixture in case hibernating animals, such as hedgehogs or frogs, have taken up home there.

Beetle

4. Your compost is ready when it looks brown, feels crumbly and smells sweet. This can take as little as six months, but you may have to wait as long as a year.

Old-timer's tip

Speed up your composting by adding some of that old gardener's favourite – human urine – into the mix (see 'Liquid feeds' on page 40).

Making leafmould

Autumn provides yet more bounty for the gardener in the form of fallen leaves. Unlike compost, which is produced by bacterial action, leafmould is the result of action by fungi and it's much simpler to make than compost. Use the leaves of deciduous trees such as oak, as these will rot down better than evergreen leaves. The easiest method is to scoop up the leaves and place them in a large bin liner. Pierce holes around the bottom of the bag for drainage. Then walk away. The following autumn, you'll have a lovely, crumbly mixture to use as a mulch (see opposite).

Green manure

If compost seems like too much effort, here's a solution involving minimum effort: grow traditional green manure. This is a crop cultivated specifically to be dug into the soil. It improves soil structure and fertility and has the extra benefit of helping to suppress weeds. Start in autumn on a bare patch of earth. Scatter the seeds over the surface, rake them in lightly, and leave them to sprout and grow over winter. In spring, simply dig the greens back into the soil. What could be easier than that? You can buy green manure seed mixes. All grow fast so help to keep weeds down and improve soil structure by adding organic matter when dug in. Some also fix nitrogen in the soil. Here are a few of the most popular plants for this purpose:

- Alfalfa
- Buckwheat
- Red clover
- White clover
- Grazing rye
- Mustard greens

Mulches

Over time soil can get depleted, so you need to top up its goodness by feeding it, improving its structure and encouraging the growth of healthy bacteria and fungi. One way to do this is with an annual mulch. This is a layer of organic matter, at least 5 cm (2 in.) thick, which is spread over the surface of the soil. It doesn't need digging in, because the earthworms will carry it down for you. Another important function of mulching is that it retains moisture – so that you don't have to water so often – as well as being a good weed suppressant.

There are a number of different mulches you can buy – bark chippings, mushroom compost, horse manure, for example. In the old days, the latter would have been available from the local stables. (If you have access to manure from a stable, make sure it has been left to rot down for two years, or it may scorch your plants.) Or you can use the mulches you have already made:

- Leafmould, although not high in nutrients, makes an excellent soil conditioner.
- Garden compost is a good all-rounder, improving soil structure and retaining moisture.

No-dig gardening

You can bury a lot of troubles digging in the dirt.

Anonymous

The saying above is certainly true – gardening can indeed take your mind off your troubles. However, there is a school of thought that says that *not* digging is better for the soil and the planet. It's even been rumoured that heavy digging was invented by head gardeners to give their young apprentices something to do! If you'd rather sit in your chair than pick up your spade, this is one for you.

The supporters of this method point out that not digging minimizes the release of CO_2 and other greenhouse gases, and sustains the billions of organisms that keep the soil healthy. By leaving the earth undisturbed and laying mulches, you are in fact mimicking what happens in nature as organic matter rots down and is carried below the surface by those wriggly helpers – earthworms.

The three essentials

For all-round healthy growth, plants need to draw up three essential nutrients from the soil. They soak up these soluble goodies through their roots, which is why they need watering.

- Nitrogen for leaves and general health
- Potash (potassium) for flowers and fruit
- Phosphorus (phosphates) for roots

Liquid feeds

Canny old-timers had various tricks for turning leftovers from the kitchen or plant material from the garden into rich liquid feeds:

Beer Pouring leftover beer on the soil can encourage stronger growth. It could be the yeast that works this magic.

Comfrey There are different ways of making this popular liquid feed, which is high in potash to encourage fruiting, and in phosphorus for strong roots. Here's an easy method. Fill a bucket (if you use a black bucket, it will absorb the sun's heat and speed up the decomposition process) with comfrey leaves; you don't need to remove the stalks. Place this, covered, in a sunny spot. In about three weeks, the soft leafy parts will have rotted away, leaving only the fibrous stalks – and a lot of extremely smelly liquid. Strain the liquid into another container, or lift out the fibrous mass with a hand fork; this can go on the compost heap to rot down further. To apply your comfrey fertilizer, dilute it by one part liquid to four parts water, and pour around your plants. Homemade comfrey fertilizer is great for fruiting plants such as tomatoes and courgettes (zucchini), and means you won't have to shell out for proprietary tomato feed.

Milk Rinse out your milk bottle and use the diluted liquid as an old-fashioned feed.

Nettles A liquid feed made from this nitrogen-rich plant encourages lush, leafy growth. Wearing gardening gloves to protect your hands, pick handfuls of nettles and press them down into a bucket. Weigh down with something heavy like a brick, cover with water and leave to rot down for two weeks. At the end of this time you will have a powerful – but very stinky – brown liquid. Dilute it by one part liquid to ten parts of water, and pour around your plants. Avoid using this potent brew on young plants, though, as it will be too strong for them.

Urine Yes, human urine – practicality must come before

personal embarrassment. A highly effective, traditional liquid feed, this natural bodily fluid is rich in nitrogen. It needs to be fresh (and avoid using it if you are on medication, because of the unpredictable effects of chemicals in the liquid). Dilute by one part of urine to twenty parts of water, before pouring on the soil around your plants.

Solid fertilizers

No old gardener would dream of throwing food waste away. It would be added to the compost heap, but some could be applied directly to the soil to feed plants.

Banana skins Dig in old skins around the roots just below the surface of the soil, with the inside of the peel facing down. They rot down quickly and are packed with all kinds of goodies that your plants will love – potassium, phosphates, magnesium, sulphur, calcium, silica and sodium.

Coffee grounds Don't throw away those coffee grounds after you've had your morning brew! Scatter them around your plants – they are packed with nitrogen, potassium and important minerals. Scattered thickly as a mulch, they also suppress weeds, so keep them away from seedlings and young plants.

Eggshells Another breakfast boost for your garden. Save the shells, rinse and dry them, then crush them finely and scatter around your plants. They contain calcium, which supports cell strength.

Wood ash The ash from log fires contains potash, so

sprinkle it around your plants. Avoid scattering it near mint or digging it into the surrounding soil, as it is said to be lethal to this herb.

> *We come from the earth, we return to the earth, and in-between we garden.*
>
> Anonymous

Across the seasons

To keep your soil in peak condition and make it easier to work, apply some old gardener's know-how.

Winter

- Cover any bare soil to keep it warmer so that you can sow earlier in spring. You'll also be suppressing weeds to save you work when the growing season starts. They may not be pretty, but old carpet or flattened cardboard boxes will do the trick, or alternatively you can buy purpose-made weed-suppressing sheets.
- Don't dig the soil when it's frozen, or you could damage its structure.

Spring

 Lightly fork around plants to give them a boost and to gently aerate the soil.

Summer

Now is the time to apply fertilizers and feeds to your growing plants.

Autumn

This is when you apply those mulches.

Back in time

How do you know when the soil is warm enough for sowing in spring? If you were a farmer in the old days, you might try the 'bum test': sit on the soil with your bare bottom. If it feels bearably cold on your skin, but not freezing, you'll be OK. You might prefer to use your elbow instead.

3

Plants, Water and Weather

All the flowers of all the tomorrows
are in the seeds of today.

Indian proverb

Whether you are a novice gardener or a seasoned old-timer, there is always more to discover about the practice of raising plants and caring for them – that's part of the joy of gardening, of course. Much of this knowledge is timeless and still applies, but some customs may sound rather quaint to modern ears. For newcomers, experienced old gardeners might as well be talking in a foreign language, so first you need to be able to translate the lingo.

The secret language of gardening

Gardening terminology includes a bewildering array of words and definitions. What does half-hardy mean? What's a herbaceous perennial? And what on earth is a rhizome? Here's a short guide to help you unpick the secret language of gardening.

An annual is a plant that germinates, flowers and dies all in one year, having dropped its seeds to start the new generation next year. Once some annuals take hold, you'll have them for ever (if you want them, that is): think nasturtiums.

A biennial takes two years to do what an annual does in one.

Bulbs are the onion-shaped, swollen stem bases of plants such as daffodils, tulips, snowdrops and hyacinths. 'Bulbous' refers to all plants with this kind of base.

Snowdrops

Corms look like slightly flattened bulbs. Crocuses, irises and gladioli grow from corms.

Tubers are elongated, swollen stems or roots – sweet potatoes are tubers. Dahlias grow from tubers.

Rhizomes are long, swollen, horizontal underground stems. Irises grow from rhizomes.

Hardy plants are tough enough to survive outdoors all year, even when it's frosty.

Half-hardy plants cannot tolerate really cold temperatures or frost.

A perennial is a long-lived plant that survives for three or more seasons.

Herbaceous perennials have soft stems and die right down to the ground in winter.

Woody perennials have woody stems and branches; although they may lose their leaves in winter, their branching framework remains.

Deciduous plants lose their leaves in winter.

Evergreen plants keep their leaves all year.

Solar-powered plants

A word about bulbs, corms, tubers and rhizomes: these little underground larders are important to a plant's survival, because they are where it packs away food for the dormant months, so that it can burst into new life the following season. This is the reason why you should not cut back or tie together the leaves of bulbous plants, such as daffodils – let them die down naturally. While green they are still doing the job of building up food stores using carbon dioxide, water and the energy of the sun. It's a natural process called photosynthesis – literally 'synthesizing light' – and is the original solar power.

Old-timer's tip

How deep should you plant your bulbs?
That's easy – two to three times as deep as
the height of the bulb itself. And don't forget
that the pointed bit should be at the top.

Growing from seed

This rule in gardening ne'er forget, to sow dry and set wet.
Traditional saying

Today, garden centres are bursting with all sorts of wonderful seedlings and plants for us to buy. It's certainly an easy and convenient way for us to stock our gardens. But just think of what goes into supplying this quick-fix produce. Seedlings are raised in pots or trays (plastic or polystyrene) in large nurseries, where they may have to be exposed to artificial lighting and heat to bring them on in time. They may be grown in compost based on peat, which is a precious resource. They are then transported in vans and lorries to individual garden centres across the country. That's a lot of energy and resources.

Experienced gardeners would have raised their own plants from seed – not to make a political point, but because a) they probably didn't have the choice, b) it saved money and c) it was just what you did! Growing your own plants from seeds does take a bit more work and needs a little extra space, but it gives you a wider choice in what you can grow, as well as enormous pride and pleasure in knowing that the flowers and vegetables before you are all the result of your own tender ministrations. You will also get a lot more for your money.

Sowing under cover

The seeds of hardy plants can be sown outside directly where they are to grow. Half-hardy types, or any that you want to get started early, need growing under cover, either in a greenhouse or inside your home – a sunny window sill is a good location (but not above a very hot radiator which could be too drying for your little protégés). You don't even need special equipment: anything that is deep enough to hold compost and give the roots enough space will do, such as old yogurt pots or plastic or foil food trays. Clean these well and don't forget to make holes in the bottom for drainage too. A heated metal skewer is a useful tool for making neat holes in plastic without splitting it.

1. Fill each container with multi-purpose compost (peat-free, of course – see page 30). Water it gently with the fine rose of a watering can and allow to drain. Watering it first means you won't dislodge the seeds after sowing.

2. Sprinkle the seeds thinly over the top and cover with another layer of compost. The golden rule is: the smaller the seeds the thinner the covering.

3. Place a sheet of glass or clear plastic over the tray or secure a clear plastic bag around each yogurt pot using a rubber band. The top half of a clear plastic drinks bottle also makes a good cloche (protective translucent dome) for pots. Doing this helps to maintain an even temperature inside and ensures that the compost stays moist.

4. Leave the containers on a window sill or other well-lit space and wait for the seedlings to appear, then remove the coverings and allow the little plants to grow on indoors until they are big enough to be transferred to larger pots, or the weather is warm enough for them to be transplanted into their final site outdoors. Hold them carefully by their leaves, not their roots, and use a kitchen fork to help you in the delicate operation of lifting them.

Old-timer's tip

Some seeds are so tiny that it's tricky to sprinkle them thinly. Try an old gardener's trick and mix them with some fine sand first.

Sowing outdoors

If you are sowing flower seeds directly where the plants will grow, it's best to use the 'broadcast' method and scatter them on the soil. They will look more natural grown this way. Take care, though, to mark out where you have sown. When the seedlings first appear you may not be able to tell the difference and think they are weeds. The little plants themselves will add to the confusion because their first leaves – their seed leaves, or 'cotyledons' – will be a nondescript rounded shape, unlike their true leaves, making it harder to determine just what they are.

You can, of course, sow seeds in rows which makes it easier to identify just what those shoots are – weeds don't grow in rows, after all. This method particularly suits vegetables, and can be really pleasing to the eye. To achieve a perfectly straight seed trench, insert an old broom handle lengthways into the soil. Sow your seeds into the groove you have made and flip the dislodged soil over them.

Another crafty trick is to fill a length of guttering with compost and sow your seeds there. When the seedlings appear, dig a little trench in the soil the same length and depth as the guttering; you can test it for size by carefully lowering the guttering into the space. Now for the really clever part: gently slide the whole row of seedlings out of the guttering and into the trench. No lifting and transplanting needed!

When sowing any seeds, it's always wise to keep a few in reserve in case some of your seedlings don't survive because of disease or because they have been munched by a snail or slug.

One for the rook, one for the crow, one to die and one to grow.

Traditional saying, referring to how many seeds to sow

Saving seeds

Nature has several ingenious ways of ensuring the continuation of plant life. One of the most common is through the distribution of seeds. Every flowering plant's aim is to replicate itself by self-seeding. After the flowerheads have dried, the seeds they contain fall to the ground, germinate, take root and, hey presto, the next generation of plants is born. From the minute black seeds of poppies to the floating seedheads of dandelions and the acorns of mighty oaks, all flowering plants proliferate in this way.

You can get in on Nature's act and save seeds from the plants you have grown yourself – and gain a whole load of new plants for free. Do this in autumn, as the seedheads (the old flowerheads) are drying. Aquilegia, cosmos and scabious are some of the flowers that lend themselves well to this conservation effort.

1.   To start with, you'll need paper bags to hold the seedheads. You can place the same species of flower together, but you'll need separate bags for other varieties. Snip off a dry seedhead and carefully lower it, head down, into a bag.

2.   Store the bag in a dry place to allow the seeds to ripen and dry out. In a couple of weeks, the heads will have dried fully and the seeds will have dropped into the bag.

3.   Label an envelope with the date and name of the flower seeds you're saving.

4.   Remove the seedheads from the bag and tap them gently over a piece of white paper to loosen any remaining seeds so that they drop onto the paper. Fold the paper in half to

make a kind of chute so that you can slide the seeds into the envelope without losing any. Carefully pour any remaining seeds from the bag into the envelope too.

5. Seal the envelope and store in an airtight tin or container in a cool, airy place, away from direct sunlight. You can even keep them in the salad drawer of the fridge. In spring, they should be ready to sow.

Old-timer's tip

If you're not sure whether your saved seeds are still viable, try this test. Sprinkle a few on some damp kitchen towel and wait to see if they germinate.

More plants for free

As well as saving seeds to grow next year's crop of annuals, there are other ways you can take advantage of Nature's drive for renewal – and get new shrubs, climbers and perennials for free – by following the time-honoured practices of:

- Taking cuttings
- Dividing
- Layering

Taking cuttings

Do this in spring or early summer, when you can take advantage of tender new growth.

1. Choose a soft green shoot and cut it just below a leaf joint or little nodule on the stem, about 5–10 cm (2–4 in.) below the tip of the stem. Remove the lower leaves and any flowers or buds.

2. Using a pencil, make a hole in a pot of compost and gently insert the base of the cutting, leaving the rest of the shoot above the surface. Water the compost and cover the pot with a clear plastic bag to maintain humidity.

3. Place in a well-lit spot but not direct sunlight, and in two to three weeks your little cutting should have taken root.

An even easier method is to place your cutting in a jar of water. Refresh the jar with room-temperature water every three or so days, and watch the roots form. You can then – gently – transplant the little plant into a pot of compost before moving to its final site.

Growing from cuttings works well with perennials such as pelargoniums (commonly known as geraniums) and chrysanthemums, as well as deciduous shrubs such as fuchsias, buddleias and hydrangeas.

Dividing

Some perennials can just get too big for the space they are in and need dividing. Dig up the plant and separate it into two or more clumps. It may feel brutal but you can slice down through the rootball with a spade. After replanting each clump, it's a good idea to cut back some of the leaves so that the new plants can concentrate their energy into growing new roots.

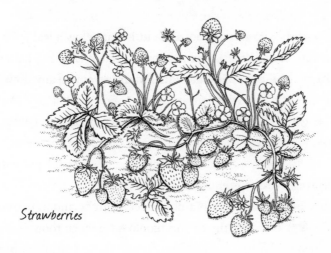

Strawberries

Layering

Some plants just can't wait to produce babies. Strawberries are a prime example and 'layer' naturally. Their enthusiasm can be hard to contain as they romp across the ground, sending out runners (long stems) in every direction and taking root wherever the little nodules on the runners touch the soil. Climbers with long, bendy stems such as clematis, honeysuckle, jasmine and wisteria or shrubs such as viburnum respond well to this method of propagation.

1. Simple layering involves bending a long, flexible shoot down to the ground and burying it in the soil at a joint, where a leaf joins the stem. You can encourage the process by making a little nick under the joint, no more than halfway through the stem. This is where the roots will form.

2. Make a hole in the soil about 5 cm (2 in.) deep, and press the joint into it so that the cut opens up a little. Pin in place with a wire loop or forked twig.

3. Tie the free end of the shoot to a bamboo cane. At this stage, the shoot is like an umbilical cord as it is still connected to the mother plant.

4. Keep the buried joint watered. Do this in autumn or spring and by the following autumn – or sooner – you should have a new plant that you can sever from its parent, and replant where you want it.

Tip layering is even easier and suits plants with long, arching stems such as blackberries and hybrids such as loganberries. Choose a suitable stem and bury the tip 7.5 cm (3 in.) deep in the soil, and peg it in place. Keep it moist and by the following autumn or spring, a new young plant will have sprouted. Sever it from the parent and move it to your chosen location.

Old-timer's tip

The basic rules for pruning shrubs are to remove dead and tangled wood to open out the middle of the plant, and cut back stems to an outward-facing bud.

Waterways

*God made rainy days so gardeners
could get the housework done.*

Anonymous

As our climate warms, water is becoming an increasingly precious commodity. In the old days, it was a scarcer resource than it is now simply because many people did not have mains water. Today we can just turn on a tap.

In *Lark Rise to Candleford*, a delightful description of a country childhood in an English village at the end of the nineteenth century, Flora Thompson (1876–1947) describes a vanished way of life. She recalls how a tarred or green-painted water butt stood against the wall of every cottage to collect rainwater from the roof. The water from this would be used judiciously on a few of the more precious plants in the garden, or for drinking, personal hygiene and washing clothes. The alternative was a trek to the well with a bucket, regardless of the weather, or to a water pump at a distant farm. Those who had their own wells guarded them fiercely with covers and padlocks.

Wells as a source of water go back centuries and, if proof were needed of their important role in everyday life, they were given special significance in folklore as thresholds between this world and that of magic.

Collecting rainwater

There is no reason why you can't emulate the old ways and fit water butts to your downpipes from the roof of your house, shed or greenhouse. They will collect the rain that falls into the gutters so that you can use it, instead of just letting it soak away. The butts are unlikely to supply all your watering needs, but it does mean that you won't have to rely totally on tap water. Some plants, such as acid-lovers, prefer rainwater.

To stop the water in the butt becoming stagnant and smelly, use it often. Also, keep your gutters clear of leaves so they can't make their way down through the pipes and into the water, where they could rot into a smelly sludge.

Wise watering

Here's some common sense know-how to help you use water wisely:

- Water in the cool of the evening when the water is less likely to evaporate. Don't water in hot sunlight as the water could scorch the leaves.
- Direct water to the base of the plants, where the roots can soak it up. Water on the leaves is wasted.
- Avoid using a sprinkler. It is incredibly wasteful.
- Water younger plants with tepid water, as really cold water can be a bit of a shock to them. If your water butt is empty, you could fill watering cans and buckets with tap water and leave them to stand for a while to take off the chill.
- Water leafy and salad vegetables especially well from three weeks before you pick them. Water fruiting crops well

when the flowers and fruit are forming. Water root crops thoroughly when the roots are beginning to swell.

 When planting out, insert a small flowerpot with drainage holes, a plastic cup with the base removed, or something similar, into the soil next to the base of each plant. You can then channel the water directly into this little conduit so it goes straight to the roots, rather than dispersing on the soil surface.

Dig plenty of organic matter into the soil so it retains moisture better – and don't forget to mulch (see page 37).

Back in time

Water is subject to fewer extremes of hot and cold than air temperature, and this can have a beneficial effect on the surrounding atmosphere. In Victorian times, head gardeners installed water tanks in their greenhouses to keep the air cooler in summer and a little less freezing in winter.

Skin care for gardeners

Soil under your nails, scratches from thorns, itchy insect bites – gardening can be tough on the skin. Here are some old-fashioned but effective remedies to sooth:

- Soil somehow has a way of working itself under your nails, even if you wear gloves. Pushing a little wet soap under the nails before you start will make it easier to clean them when you have finished.

- To remove stubborn green stains on your hands, cut an unripe tomato in half and rub it on the stains before washing your hands.

- Insect bites are an occupational hazard for gardeners, especially at times when midges and mosquitoes are on the prowl. Immediate relief may be found by rubbing the bites with a marigold leaf.

- Dabbing a bite with neat white vinegar is said to relieve itching, but don't use on broken skin.

- Apple cider vinegar, rubbed on the skin, is a homespun insect repellent. Insects will be put off by the smell (but then so may you – who wants to walk around smelling like a salad?!).

Weather lore

Knowing what the weather will bring is crucial for gardeners, for it determines such important decisions as when to plant out, how much to water, etc. – or, indeed, whether it will be too wet to get out in the garden at all. Today we have meteorologists using satellites and all sorts of fancy scientific resources to predict the weather for us. In the old days, people had to rely only on personal observation, so they kept a close eye on the natural world for signs of what was to come. The observations of these old weather watchers are encapsulated in a huge repository of old country wisdom and ancient homilies, sayings and rhymes.

We still use a few of these sayings today, and some even make scientific, or at least common, sense. Take 'Clear moon, frost soon', for example. It stands to reason that frost is more likely on a clear, cold night than when a thermal layer of cloud blankets the sky. Then there is the famous 'Red sky at night, shepherd's delight'. Meteorologists tell us that high pressure traps dust in the air, dispersing blue light and leaving red, and it's high pressure that also brings good weather.

Animals often feature in the proverbs as if, being wild creatures and thus closer to Nature, they have a kind of sixth sense when it comes to the weather and climatic events. There may be some truth in this, for animals can be sensitive to changes in atmospheric pressure. There are also many reports of animals behaving strangely before earthquakes, and before other natural disasters.

This old folk wisdom is not confined to the English-speaking world, of course; counterparts are found in other languages, too. For fun, test some out to see if they are accurate.

Sky and earth

Red sky at night, shepherd's delight.
Red sky at morning, shepherd's warning.

Rain before seven, fine by eleven.

Clear moon, frost soon.

When halo rings the Moon or Sun, rain's approaching on the run.

Mackerel sky, mackerel sky, never long wet, never long dry.
(A 'mackerel sky' is covered with fleecy clouds like the markings on a mackerel's skin, and signifies a change in weather.)

If clouds move against the wind, rain will follow.

Dew on the grass, no rain will come to pass.

Pine cones open up when good weather is coming.
(Dry weather causes the scales to dry out and fan outwards.)

Feast days and seasons

If Candlemas day be sunny and bright, winter will have another flight;
if Candlemas day be cloudy with rain, winter is gone, and won't come again.
(In the Christian calendar, Candlemas falls on 2 February and commemorates the presentation of the infant Jesus at the temple. In Celtic tradition, 1 February is St Brigid's Day and also the old festival of Imbolc, the beginning of the Celtic spring.)

Quand il pleut pour la Chandeleur, il pleut pendant quarante jours.
(A French version that translates as: 'If it rains on Candlemas Day, it will rain for another forty days.')

So many mists in March, so many frosts in May.

When March blows its horn, your barn will be filled with hay and corn.
(Here 'horn' means 'thunderstorm', and a March storm signifies unusually warm weather at this time of year.)

April showers bring forth May flowers.

Ne'er cast a clout till May be out.
('Clout' means item of clothing, which you should not 'cast off' until 'May be out'. Here, 'May' can be interpreted as the month of May, or as 'may' which is another name for the hawthorn that flowers around this time.)

Shallots should be planted on the shortest day of the year and harvested on the longest.

St Swithin's day if thou dost rain,
For forty days it will remain.
St Swithin's day if thou be fair,
For forty days 'twill rain nae mare.
(St Swithin was a ninth-century English bishop whose feast day falls on 15 July. According to an English tradition, apples should not be picked before St Swithin's Day.)

Birds and animals

A cow with its tail to the west makes the weather best,
A cow with its tail to the east makes the weather least.
(This British proverb alludes to the westerly winds that bring good weather, and easterly winds that bring bad weather.)

When cows are lying down in a field, rain is on its way.

If crows fly low, winds going to blow.
If crows fly high, winds going to die.

When swallows fly low, rain is on
the way.

When the goose flies high, fair
weather. If the goose flies low,
foul weather.

When the bees crowd out of
their hive,
the weather makes it good to
be alive.
When the bees crowd into their
hive again,
it is a sign of thunder and
of rain.

When ladybirds swarm, expect a day
that's warm.

Spiders leave their webs when it is going to rain.

4

The Herb Garden

The garden is the poor man's apothecary.

Traditional German saying

Growing herbs is a practice that goes back many centuries. One of the earliest-known herb gardens was created nearly three thousand years ago in Babylon, by King Mardukapal-Iddina II. It included sixty-four species, some of which are still familiar today, such as dill and fennel.

For our ancestors, however, 'herbs' were much more loosely defined than they are today. What they called a herb, we might now label 'flower', 'vegetable' or even 'weed'. Thus, alongside sage, camomile or fennel, they might have grown pot marigold, lavender and yarrow – which we would associate with ornamental gardens – or even groundsel or chickweed, which to us are weeds.

What defined a herb was not so much how it *looked* as what it could *do*. More than anything, herbs were plants-with-a-purpose that were grown for their practical application as medicinal or culinary aids. Hence the old name for a dedicated herb garden: the 'physic garden' – the garden of the physician.

The story of herbs takes us on a horticultural journey through time, from the Middle Ages through to the nineteenth century and beyond.

The medieval medic

What can be more pleasant to thee, than the enjoying of medicines for cure of thine infirmities, out of thy native soil, and country, thy field, thy orchard, thy garden?

Nicholas Culpeper (1616–54), herbalist

Before the availability of over-the-counter medicines or prescription drugs, herbs were the go-to source for cures for a myriad of ailments. They also offered a way to add extra flavour to food and dispel unpleasant odours at a time when modern sanitation or hygiene practices had yet to be developed. The formal physic gardens in which these wonder-plants were cultivated were found in the grounds of medieval monasteries, for back then it was the job of the monks to tend to the sick and infirm. A garden of this kind was known as a *herbularis* or *hortus medicus* (literally 'medicinal garden' in Latin). Herbs and medicines were kept in a storeroom called the *officina*, echoes of which can still be found in the word *officinalis* that forms part of the botanical names of some herbs, such as *Salvia officinalis*.

Book learning

He who has sage in his garden will not die.

Traditional Arabic saying

The monks drew their knowledge of healing herbs from ancient Classical and Arabic texts, which they translated. One of the oldest such herbals was the *Lacnunga*, dating from the tenth and eleventh centuries, although some of the material it contains is even more ancient. Mostly written in Old English and Latin, it takes its name from the Old English *laec*, or 'healing', and includes around two hundred herbal remedies. The preparation of some bordered on magical ritual and required the herbalist to recite a prayer or invocation while the remedy was made. One example from the *Lacnunga* was the Nine Herbs Prayer, to be intoned while concocting a medicinal salve for a particular skin inflammation, made from mugwort, plantain, shepherd's purse, nettle, betony, camomile, crab apple, chervil and fennel.

Several centuries later, two very famous herbal guides appeared on the scene. The first was John Gerard's *Herball, or General Historie of Plantes*, published in 1597. Gerard was a botanist with a large herbal garden in London, and in his great work he detailed the names, habits and 'vertues', or uses, of more than one thousand plants.

Around fifty years later, what is perhaps the most comprehensive – and almost certainly the most famous – book on herbs was published. This was the *Complete Herbal* of 1653, written by Nicholas Culpeper, an English herbalist, physician and astrologer.

In the monastery garden

Here are just a handful of the herbs that would have been grown in a medieval monastic physic garden, along with the properties ascribed to them:

Betony This was truly a wonder herb for tradition claimed it could treat just about everything, from kidney stones, flatulence, coughs and lung problems to deafness, 'specks on the eyes', poor eyesight and – if taken regularly with wine – improve a pasty, dull complexion. Moreover, it was reputed to protect against 'fearful night visions', and would even deter snakes.

Camomile Familiar to us for its calming properties, this was said to relieve flatulence and digestive problems, while an infusion of camomile and other herbs was a favoured remedy against poisoning.

Clary sage Known as 'clear eye' or 'eye of Christ' (*Oculus Christi*), this was used to make an eyewash.

Camomile

Comfrey This was valued for its power to heal wounds, soothe inflammation and set bones – hence its common name: 'boneset'.

Cumin The seeds were used in ointments to soothe the skin and eyes. Peasants living outside the monasteries sometimes paid part of their rent in this herb.

Hyssop This was seen as a purgative that could rid the body of ailments such as catarrh and phlegm, and as a poultice for burns and bruises. Also known as 'holy herb', it was used by Benedictine monks to flavour the liqueurs they made.

Rue This pungent herb was employed not only as a powerful purgative in cases of plague or poisoning, but was also used in exorcisms.

Sage The 'salvia' part of its name comes from the Latin *salveo*, or 'I am well', as found in such words as salve and salvation. It was said to be 'fresh and green' and able to 'cleanse the body of venom and pestilence'. Sage was also chewed as a handy tooth whitener.

Back in time

Used in spells and incantations, yarrow was known as the witch's herb. It is also said to be effective in staunching bleeding and helping wounds to heal, and was used in this way by soldiers in World War I. Going back many centuries to another war – the Trojan – the Greek warrior-hero Achilles was reputed to have used it to treat his soldiers, hence its botanical name *Achillea millefolium*.

No nasty odours

Queene Elizabeth of famous memory, did more desire it
[meadowsweet] *than any other herb to strew her chambers withall.*

John Gerard, *Herball* (1597)

In France, the warning call was *garde a l'eau* – in Britain, 'gardy-loo' – as the contents of chamber pots were thrown into the street. With smelly rivers, no flushing toilets or sewers to carry away waste, bathtime a rarity, and no fridges to keep food fresh, people and places could all get a bit stinky in centuries past. From the Middle Ages right through into the early nineteenth century, herbs again offered a solution.

These aromatic plants were used to preserve meat and to mask the smell of rotting food and body odours. 'Strewing herbs' such as tansy, wormwood, wall germander and meadowsweet were also scattered, or strewn, over floors; when trodden on, they released their fragrance and helped to cover up less desirable aromas. They could repel insects, too.

The folk herbalist

The plant voideth away the worms, not only taken
inwardly, but applied outwardly; it withstandeth all
putrefactions, and is good against the stinking breath.

John Gerard, describing the properties of wormwood in *Herball* (1597)

Sweet herb pot pourri

For a traditional pot pourri to perfume your home, you will need the dried leaves of some scented herbs such as rosemary, bergamot, mint, thyme and bay (dry them gently in an airing cupboard or a cool oven with the door open). Arrange in layers in a large jar, sprinkled with sea salt, ground cloves, ground cinnamon, lavender flowers and dried tangerine peel. Leave the mixture for a few weeks, stirring occasionally. When the scents have blended together, place the mixture in a porcelain, earthenware or glass jar, and allow the fragrance to waft around the room. To refresh the pot pourri, stir in a few drops of an aromatic essential oil, such as lavender.

Medicinal herbs were not found only in the formal physic gardens of learned monks. Ordinary people had acquired knowledge of them, too, and grew them in their own humble plots to make their own folk remedies. The age-old knowledge of herbs was passed on down the generations and when the first European settlers began to arrive in America from the seventeenth century onwards, they brought their herbs with them too. They grew them close to their homesteads so they were readily to hand, and used them to treat illnesses, strew on floors, perfume linens, flavour meat and dye fabrics.

Back in the Old World, people continued to grow herbs

and would have had an encyclopedic knowledge of their many applications. This is perhaps more the preserve of the 'Old Wife' than the 'Old Boy', for in the domestic sphere it was the women rather than the men who were responsible for feeding and caring for their families. Thus an Old Wife, adopting her time-honoured role as Healer and Wise Woman, might concoct herbal remedies to soothe sore throats, aid digestion, ease headaches, foster sleep, repel insects, add lustre to hair, improve the complexion, flavour food, mask nasty smells and more.

In a country garden

The common names of these herbs reveal how they were perceived, and some were applied according to the principles of 'sympathetic medicine': in other words, if a plant bore a resemblance, in shape or colour, to the symptoms of a particular disease, this signified it could cure that disease.

Alecost or costmary An infusion of the leaves was used to treat colds, catarrh and upset stomachs. Its camphor fragrance also made it a good insect repellent, so it was inserted within the pages of the family Bible to keep it free of moth damage, hence its common name: 'Bible leaf'.

Cowslip When the keys to Heaven dropped to the ground from St Peter's belt, they were instantly changed into these golden-yellow flowers, which is why they are also known as 'keys of St Peter'. Cowslips were a remedy for headaches, and were added to salads, stuffings and wine.

Cowslip

Fennel or 'finkle' One of the oldest plants in cultivation, this had both culinary and medicinal applications. Chewing fennel seeds helped to freshen the breath.

Figwort The little swellings on the roots were thought to resemble figs, and *ficus* (Latin for 'fig') was also an old name for haemorrhoids, or piles, making figwort the obvious remedy for this unpleasant condition. The plant belongs to the genus named *Scrophularia*, as it was also used for scrofula: swelling of the lymph nodes in the neck.

Good King Henry Used to clean and heal sores on the skin. The young shoots were eaten as a substitute for the asparagus enjoyed by wealthier people, hence its nickname: 'poor man's asparagus'.

Ground ivy This evergreen creeper provided a useful cough treatment. As a flavouring for ale, it was known as 'alehoof'.

Horehound or soldier's tea A member of the mint family, this was a popular cough remedy for thousands of years. As an extra benefit, it was also thought to protect against magic spells.

Orange hawkweed The clusters of unopened buds beneath the open flowers reminded people of fox cubs sheltering beneath their mother, so they nicknamed it 'fox and cubs'. It was used for digestive complaints and flatulence.

Tansy The tight little yellow flowers of this plant probably gave rise to its common name, 'bachelor's button'. It was used to repel flies and other insects.

Wall germander or ground oak A remedy for dropsy, jaundice and gout; in powder form it was used as snuff and to relieve head colds.

Wormwood This repelled both insects and evil spirits. It was associated with the spiritual rituals on St John's Eve (23 June, the eve of Midsummer Day during the Summer Solstice period), which gave rise to one of its other names, 'St John's girdle'.

Back in time

Botanically, bitter-tasting wormwood is known as *Artemisia absinthium*. It was used in the brewing of absinthe, which Van Gogh is said to have been drinking before he sliced off his ear, possibly due to its alleged mood-altering properties, although these have been disputed.

Be your own herbalist

Growing your own herbs is a joy, for these are such multi-purpose plants. If you are lucky enough to have the space, you can create your own formal herb bed, perhaps edged with low hedges of lavender. But you can equally well grow herbs in containers, or dotted among ornamental flowering plants or vegetable crops. If you choose the latter, herbs can also act as 'companion plants' to attract pollinators and deter pests (for more on this, see the Pests and Diseases chapter).

For greatest growing success, provide your herbs with the kind of conditions that most closely resemble those where the plants originated.

> *I like muddling things up; and if a herb looks nice in a border, then why not grow it there? Why not grow anything anywhere so long as it looks right where it is? That is, surely, the art of gardening.*
>
> Vita Sackville-West (1892–1962), British garden designer and writer

Back in time

Lavender has an ancient pedigree. Its botanical name, *Lavandula*, comes from the Latin *lavare*, meaning to wash, because the Romans often used lavender oil in the cleaning of clothes.

Herbs for shade

A shady spot is preferred by some herbs because conditions here are more to their liking – cooler air and nice, moist soil, which will help to prevent them bolting (shooting up and going to seed). While deep shade will turn them straggly as they reach for the light, they will tolerate dappled shade, or somewhere that receives some sun during the day. Angelica, chervil, coriander (cilantro), dill, lemon balm, lovage, mint, mustard, parsley, rocket (arugula) and sorrel are among these shade-tolerant plants. A word about mint: this spreads by the cunning method of underground runners, so if you don't want your herb patch overrun, keep it contained in a pot – there is only so much mint tea anyone can drink.

Parsley seed goes nine times to the Devil.

Traditional saying (meaning that it takes a long time to germinate)

Keeping in shape

Left to its own devices, lavender – like its Mediterranean companion, rosemary – can end up as a mass of leggy, woody stems with very few leaves on top. To prevent this happening, follow the old practice of getting out your shears after the flowers have died and giving the plant a good crew-cut, cutting across it halfway down the stems. This encourages new growth lower down. Take care, though, not to cut away all the green growth or you may end up with a tangle of dry sticks and a dead plant!

Lavender

Harvesting herbs

The botanical structure of a herb determines the best way to pick it, and also helps to preserve the life of the plant and keep it growing.

- Herbs with leaves along the stems, such as mint, bay and thyme, should be cut where the leaves meet the stems. Herbs with long stems growing up from the base of the plant, such as parsley and chives, should be cut near the base with scissors or a knife.
- Soft-leaved herbs, such as basil or parsley, should be picked before the flowers form. After this, the plant will be focusing all its energy on producing flowers and seeds, and will be past its best.

Lavender sugar

Stir 2 teaspoons of lavender flowers into 1 kg (2 lb) of caster sugar, and store in jars. Use in shortbread, sprinkled over sponge cakes or berries, or give as gifts. Rosemary sugar can be made in the same way.

The herbal medicine cabinet

If you've run out of painkillers or any other everyday medication, try some of these old-fashioned remedies to relieve the problem:

Tension headaches: rub your temples with mint leaves.

Mint

Bad breath: chew parsley or mint.

Gingivitis (gum disease): chewing mint leaves will relieve the discomfort – but also get yourself to a dentist double-quick!

Mouth ulcers: to ease the pain, place a basil leaf on the spot and leave it there for as long as possible.

Sore throats: drink an infusion of rosemary. It's anti-inflammatory and antibacterial. It is also said to boost the immune system.

To stimulate the flow of breast milk: drink fennel tea. (An infusion of fennel seeds is also a traditional herbalist's remedy for colic in babies.)

To aid digestion: drink mint tea. The menthol in it soothes the lining of the digestive tract and stimulates the production of bile, which is essential for good digestion. A fennel salad is also an excellent way to finish a meal for good digestion.

For calmness and serenity: drink camomile tea.

Herbal beauty treatments

For blondes: infuse half a cup of camomile flowers in boiling water for an hour or so. Strain and use the camomile water to rinse your hair and lighten it.

For brunettes: infuse rosemary leaves in the same way and rinse dark hair with the infusion to add extra lustre.

For oily skin: infuse some dried yarrow flowers in a large cup of boiling water and leave to steep. When cool, strain the liquid and dab on the skin. Yarrow has astringent properties.

Homemade soap

The herb saporia, or soapwort, was once used to make soap. You can make your own old-fashioned and beautifully perfumed soap with some dried, crushed lavender flowers and a few drops of lavender oil. Grate some good-quality, unscented soap into a bowl with a cup of boiling water. Place the bowl in a saucepan of water over a low heat, and stir until the mixture is smooth. Remove the bowl from the heat, stir in the dried lavender and oil. When cool enough to handle, mould into shape with your hands and leave to dry on greaseproof paper.

Herbal insect repellents

Have you ever sprayed your skin with a preparatory insect repellent and found yourself coughing and spluttering as the powerful chemical vapour hits your nostrils? If so, there is another way. Old-fashioned herbal repellents are gentler on the human nose, but off-putting to the buzzing and biting pests that are annoying you.

To deter flies from entering your kitchen: grow basil or mint in a pot near the kitchen door.

To keep flies away from food: fill some cushions with dried mint or place bunches of fresh mint near where you keep the food.

To stop flies buzzing around you when you are outside on a hot day: rub your face and neck with mint leaves. An infusion of camomile, applied to the skin, has a similar effect.

To keep midges from feasting on you: wear a bunch of sage leaves around your neck.

To repel fleas and ticks from horses: a traditional method was to scatter fresh herbs in the stables.

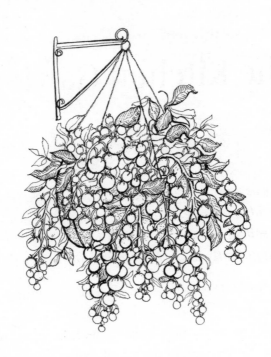

5

The Kitchen Garden

It was one of the most bewitching sights in the world to observe a hill of beans thrusting aside the soil, or a row of early peas just peeping forth sufficiently to trace a line of delicate green.

Nathaniel Hawthorne, *Mosses from an Old Manse* (1846)

Edible plants would have been at the heart of an old-time garden. Flowers delight the eye, of course, but a garden was too important a resource to assign purely to ornamental plants: it had to be productive and provide food, too. This isn't to say that a kitchen garden can't be beautiful. One of the most attractive small gardens at London's famous Chelsea Flower Show one year was devoted entirely to vegetables: lush cabbages as opulent as any rose; stately, architectural sweetcorn; frothing broccoli; luxuriant potato foliage; round, ripe squashes…

Cabbage

The vegetable patch

There is something especially wonderful about growing your own vegetables and fruit: something indefinable and perhaps even more innately pleasing than having an ornamental garden. It has something to do with the deep satisfaction of producing your own food, a deeper connection with the earth, with Nature, with our roots. And, of course, nothing beats the flavour of home-grown food.

The potager

Today, it's so easy for us to pop out to a supermarket or local store to buy our fruit and vegetables, produced for us by others. What is so hard for us to grasp is that once 'growing your own' was not just an enjoyable bit of fun but was often an absolute necessity: if you didn't grow it, you couldn't eat it. Country dwellers had an advantage here over city inhabitants, for they had the space to cultivate their own vegetables and fruit to feed their families.

Living communally and working together for the benefit of the whole, monasteries were ahead on this one. In medieval France, monks created a particularly beautiful type of kitchen garden called a 'potager'. The name comes from the French word *potage*, which is a thick soup. They divided up their potagers into four intersecting beds in the shape of a cross, to evoke the cross on which Christ died. There, they grew all the fresh food they needed.

Later, the idea was taken up by the French aristocracy, who established grand kitchen gardens and parterres based on a similar grid pattern. Perhaps the most exorbitant was created for the French Sun King, Louis XIV, next to his palace at Versailles. Known as the *Potager du Roi* (the King's kitchen garden), it was located on a marshy site, covered around 9 hectares (22 acres) and took five years to complete. Ever the hedonist, Louis had his gardeners grow luxury, out-of-season crops, such as asparagus in January and strawberries in March. This royal kitchen garden still exists and is open to the public.

Back in time

Thomas Jefferson, chief author of the Declaration of Independence in 1776 and third American president from 1801 to 1809, was a keen gentleman farmer. At his estate in Monticello, Virginia, he liked to experiment with different crops, such as broccoli imported from Italy or figs from France.

The allotment

*Every house had a good vegetable garden
and there were allotments for all...*

Flora Thompson (1876–1947), *Lark Rise to Candleford*

The tradition of the humble allotment – parcels of land made available for people to grow their own fruit and vegetables – goes back centuries. Not only did these little kitchen gardens enable people to supplement their diets, but they also had physical, psychological and social benefits. The produce and outdoor exercise involved in growing it were good for health, offered an enjoyable hobby, gave people a sense of purpose and helped to foster a sense of community. No wonder there are often such long waiting lists for them today.

Allotments really came into their own in the two World Wars, when they became a necessary part of the war effort. With food in short supply and rationing in place, governments encouraged people to give every possible open space over to growing crops. Even private gardens and parks were turned into allotments, which were now known as Victory Gardens. In Britain in World War II, the government pressed the message home with a highly successful slogan: 'Dig for Victory'.

Back in time

Onion juice has antibiotic properties. In the American Civil War, it was used to treat gunshot wounds. 'I will not move my troops without onions,' declared General Ulysses S. Grant, leader of the victorious Union Army and later US president. Garlic juice has similar antibiotic and antiseptic qualities, and was used as an antiseptic in both World Wars. Onion juice is also an old-time home remedy for baldness.

Doing the rounds

Candlemas day, put beans in the clay.

Traditional saying

Even if the kitchen garden was not laid out like a formal potager, some rules still applied. Every old-time gardener would have known and practised – to some extent anyway – the time-honoured system of 'crop rotation'. This works a bit like a circular queueing system, with individual groups of vegetables waiting in turn to grow on the bit of ground where a different group grew the previous year. The system has been used in kitchen gardens and, on a much larger scale, in farmer's fields for a very long time. The purpose is to avoid a concentration of pests and diseases specific to a particular group of plants building up in any one area. It also allows the kitchen gardener to supply the different crops with the exact conditions they like.

Of course it isn't totally straightforward, as different gardeners may place the same vegetables in different groups. Grouping also depends on how many years rotation will last: four years needs four vegetable groups, three years needs three. Here's a good, workable example for a four-year plan:

Potatoes – these need a moist, rich soil, so add compost if necessary

Root crops – these include onions, leeks, carrots, parsnips and beetroot which need a free-draining soil, so add coarse grit or sand if your soil is heavy

Legumes – these include runner beans, French beans, broad beans, peas

Brassicas – these include cabbages, cauliflowers, Brussels sprouts, swedes (rutabagas), turnips and radishes (yes, these last three belong here and not with root vegetables), and need a reasonably well-drained soil

Add to these four key groups a fifth made up of vegetables that aren't fussy, such as lettuces, tomatoes and courgettes (zucchini), which can be slotted in wherever you have space. You may not have room to follow the crop rotation system to the letter, but you can at least alternate where you grow the different groups each year.

Thirsty vegetables, such as runner beans and celery, need to be well hydrated. Dig a trench for them before planting and fill it with a layer of moisture-retaining material, such as old newspapers.

When sowing your vegetables, remember to walk on a plank between the rows in order to avoid compacting the soil.

Back in time

From a handful of wild tubers, the Incas are said to
have developed more than three thousand varieties
of potato, which they grew in the valleys of the Andes.
When this tuber was first brought to the Old World,
Europeans were less than impressed. The Spanish
saw it as food fit only for slaves, the Germans fed it to
prisoners and animals, while others suspected it of
causing leprosy because its lumpy shape and pock-
marked skin resembled the symptoms of the disease.

Perfect planning

Here are two canny old ways of getting the most from the space
you have. It's all to do with timing:

Intercropping involves growing early crops next to or
in-between rows of later-maturing vegetables. The early
ones will be ready to harvest when the later ones are just
beginning to mature. Check out the sowing and growing
times of different crops and you can make a start.

Successional sowing just keeps 'em coming. Grow quick-
maturing plants such as salad vegetables and carrots at
regular intervals – say, every two weeks during spring and
summer – to have a regular supply.

The fruit garden

A cherry year, a merry year.

Traditional saying

Berry fruits can crop away happily on their own, but stone fruits such as apples or cherries have traditionally needed a partner – a different variety of the same tree but which flowers at the same time. Nectar-seeking insects collect pollen from one tree and transfer it to the blossom of the other, in a process called 'cross-pollination', which ensures fruiting.

At least, that's how it was back in the old days, when gardeners had to have two trees to cross-pollinate if they wanted a decent crop. Today, however, clever plant breeders have developed super-small, self-fertile varieties of fruit tree that can be grown on their own in space-hungry modern gardens. The downside is that the varieties of fruit on offer are more limited than with traditional cultivars.

Ancient apples

An apple a day keeps the doctor away.

Traditional saying

Of all fruit trees, the apple is perhaps the most iconic and romantic. No other fruit is so steeped in myth and legend: there is the apple eaten by Eve in the Garden of Eden; the golden apple given by Paris to the goddess Aphrodite that triggered the whole Trojan war; the Island of Apples (Avalon) to which King Arthur was taken when he died; and the poisoned apple out of which Snow White took a fateful bite.

Apple tree

Cultivated apple trees are the descendants of the wild form that probably originated in Central Asia. The fruit has a history going back more than eight thousand years. Transported to Europe along the Silk Road – as the trade routes from China across Asia and into southern Europe were called – the fruit was first cultivated by the Romans, using methods that we still practise today. Over the centuries, people have developed thousands of apple varieties, with their own distinct flavour and aroma. Sadly, many of these heritage fruits are now rare. However, it is still possible to buy heritage apples from specialist growers. The apple that, according to legend, fell on Isaac Newton's head and inspired his theory of gravity was a variety known as Flower of Kent, which is still in cultivation today. The apple variety known as Hunthouse, also still available, was reputedly taken by Captain Cook on his voyages of discovery to help prevent scurvy among the crew.

If you have space in your garden, why not grow the apples that would once have graced the orchards of yore? You can buy trees grown on dwarfing rootstocks so that they won't grow too large.

Heeling in

Specialist nurseries often supply fruit trees as bare-rooted saplings. To make sure that the exposed roots do not dry out, keep them soaking in a bucket of water while you prepare the planting hole. Ideally, you will plant out the young tree as soon as it arrives. If this isn't possible, you can 'heel in' the tree by burying the roots in a temporary hole, firming the soil around them, until you are ready to move it to its permanent position.

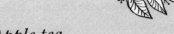

Apple tea

Refreshing, old-fashioned apple tea couldn't be easier to make. Place some sliced, unpeeled apples in a saucepan, cover with water and simmer gently for an hour, then strain. Drink this tea hot or cold.

Blessing the trees

In the cider-producing areas of England, an old ritual called 'wassailing the orchards' was carried out on Twelfth Night. 'Wassail' comes from the Anglo-Saxon toast *wes hál*, or 'be hale', and the ceremony involved blessing the trees to induce them to produce bountiful crops the following year. In the darkness of night, people headed out to the orchard, carrying jugs of cider, to toast the health of the largest tree. They poured cider on the roots, and pieces of bread soaked in cider were placed in the branches. Every magical ritual must have its own spell-song, and there were several different wassailing songs, like the one below. A similar custom is traditional in Normandy too, where the apple trees produce the raw material for the region's famous Calvados, or cider brandy.

> *Here's to thee, old apple tree,*
> *That blooms well, bears well.*
> *Hats full, caps full,*
> *Three bushel bags full,*
> *And all under one tree.*
> *Hurrah! Hurrah!*
>
> Traditional wassailing song

Berry fruits

Strawberries are another must-have in the fruit garden. They are easy to grow – almost too easy; the little plants are so determined to proliferate that they send out runners in every direction to take root wherever they can. Experienced gardeners trim off the runners as they appear. Removing the runners has a practical purpose because it helps direct the plant's energy into producing fruit. To improve growth and taste, old-timers might mix pine needles in with the traditional straw mulch. At the end of three years, maintain the time-honoured practice of discarding old strawberry plants and replacing them with new ones, grown from runners.

Raspberries are another easy-to-grow berry, being tolerant of very different climates, from very cold to hot. Because these grow on shallow-rooting canes, make sure the ground is free of weeds before planting, as vigorous weeding later on could damage the roots.

For happy, vigorous blackcurrants, follow the old gardener's lore of growing them in an old nettle patch, or plant nettles among established plants. Nettles are a little picky about the soil in which they grow, and need one that is rich in phosphates and nitrogen. If nettles are thriving it's a sign that the soil is nutrient-rich, which in turn will benefit your blackcurrants too.

Strawberry cider

Try this gorgeous, old-fashioned strawberry tipple on hot summer days. You'll need 500 g (1 lb) of strawberries, 2 tablespoons of caster sugar, the juice of 1 large orange, plus cider, to fill a large jug. Slightly crush the strawberries, sprinkle with the sugar and the orange juice, and leave the mixture to steep for 1 hour. Transfer to the jug, top up with the cider and ice, if liked, and serve at once.

Fruity remedies

- Raspberry leaf tea is an age-old treatment for sore throats and diarrhoea.
- Strawberries, cut in half and rubbed on the teeth, were once used as a tooth whitener.
- In medieval times, a strawberry purée was served to newly married couples for breakfast, because the berries were thought to be an aphrodisiac.
- Blackberries were allowed to grow freely in medieval churchyards in the belief that they would stop the spirits of the dead rising from their coffins.
- Rhubarb leaves are highly toxic because of their oxalic acid content, but the dried roots are a traditional remedy against intestinal disorders.
- Apples, apricots, cherries and peaches all have seeds that contain cyanide, but fear not: the levels are so small that you will still live if you swallow any seeds.

Bringing in the harvest

September blow soft, till the fruit's in the loft.

Traditional saying

Now for the really fun bit: harvesting the crops you have grown. But it's not a matter of just picking them whenever you want: a little good old know-how will help you to identify the moment when certain crops are 'ripe and ready'.

To tell when apples and pears are ready to pick, cup the fruit in your hand and gently twist the stem. If they come away from the branch, they are ready to pick. Use the same technique for peaches but handle them very carefully to avoid bruising the fruit.

To retain the intense flavour of tomatoes, pick them before watering, because water dilutes their fruity flavour. To retain the crispness of salad crops, harvest a few hours after watering to allow the water to hydrate the leaves. Lift potatoes in dry weather and leave them on the ground for a couple of hours before using or storing.

When harvesting peas and beans, remember that these legumes have special properties. They absorb nitrogen – an essential plant nutrient – from the air, which they fix into the little nodules on their roots, thus making their own fertilizer. When they have cropped, don't dig them up completely but leave the roots in the soil to continue feeding it.

The first Thanksgiving

The great American celebration of Thanksgiving, held on the fourth Thursday of November, traces its origins back to the seventeenth century. On 16 September 1620, around one hundred English passengers set sail from Plymouth on the English coast. Their ship was called the *Mayflower* and they were bound for the New World, where they hoped they could start a new life, with greater religious freedoms.

After a rough, sixty-six-day crossing, these 'Pilgrims', as they were known, dropped anchor and finally settled in what is now Massachusetts, in a place they called 'Plimouth' after the port from which they had sailed. Their first winter was freezing and nearly half of them died of disease and malnutrition. In spring 1621, however, rescue came from an unexpected source: the Native Americans who already lived there. They showed the newcomers how to plant 'Indian corn' (maize or sweetcorn), draw sap from maple trees, catch fish in the streams and trap beaver. By autumn of that year, the Pilgrims were able to celebrate their first harvest, accompanied by about ninety Native Americans who joined in the festivities with them. The occasion is recalled in these beautiful words written by a Pilgrim called Edward Winslow in 1621:

> *God be praised we had a good increase… Our harvest being gotten in, our governor sent four men on fowling, that so we might after a special manner rejoice together after we had gathered the fruit of our labors.*

Storing your harvest

Eat leeks in March and garlic in May,
and all the year the doctors play.
Traditional saying

Nothing beats freshly picked crops. Some need eating fairly soon after harvesting but frugal old-time gardeners would have stored others for the winter, in a cool, dark, airy place or frost-proof shed.

Here are some traditional ways of keeping fruit and vegetables in good condition for the darker months:

- Apples and pears – but only the unblemished fruits – can be wrapped individually in greaseproof paper and arranged in single layers in slatted wooden boxes to allow air to circulate.
- Potatoes need to be kept away from the light to prevent them turning green (this indicates high levels of the toxin solanine, so it's safest to throw such potatoes away). Store them in sacks or in a wooden box lined with newspaper or straw.
- Beetroots and carrots can be stored in a wooden box, layered alternately with slightly moist sand or soil.
- Onions and shallots will keep if strung together and hung up in a cool, airy place.
- Squashes, pumpkins and marrows are easy – store them on a shelf in a warm place.

Clamping

This is the name for an old way of storing root vegetables. Choose a dry site where rainwater won't collect, and dig a shallow pit about 1.2 m (4 ft) wide. Fill it with a 15-cm (6-in.) layer of fluffy straw, pile the vegetables on top, and cover with another layer of straw of the same thickness. Cover the top and sides of the mound with a 15-cm (6-in.) layer of soil, leaving a little 'chimney' of straw poking through for ventilation. And there you have it: your own vegetable clamp that any old gardener would have been proud of. Snug inside, the roots should keep for several months.

The hive and the coop

Chickens clucking in the yard and pecking at the ground, bees humming in the flowers… some small-scale livestock would have formed an essential part of the classic old-timer's garden. Along with fruit, vegetables and herbs, these would have met most of the kitchen gardener's culinary and dietary needs, providing (with apologies to vegetarians and vegans) honey, eggs and even meat.

Keeping chickens

The Country Housewife's Handbook, first published in 1939, has a lot to say on the subject of keeping chickens. Dedicated to 'country women throughout the world', it is a compendium of no-nonsense advice, instructions and knowledge gleaned from the West Kent Federation of Women's Institutes, that formidable band of women who knew everything about running the house and managing the garden. The Country Housewives advise that it is 'best to err on the high side' when deciding how many birds to have. Better, they say, to have twice as many hens as there are members of the household to avoid shortages in the winter when egg prices are highest.

They go on to state what age the pullets (young hens) should be when you buy them – sixteen to eighteen weeks old to give them time to settle down before they start laying – and describe what kind of henhouses they should have – well-insulated to keep the birds warm and with a dry floor to prevent disease. More advice follows in the *Handbook* on what to feed the birds and when, how to maintain hygiene, why you should use artificial light in winter to make up a seventeen-hour day, and how to hatch your own chicks, if that is what you choose to do.

Old-timer's tip

With all those eggs that your hens are producing, you'll need to know which are still safe to eat. Before the invention of fridges, people had other ways of testing for freshness. Here's how… Lower an egg into a glass of water. A fresh egg will sink to the bottom. An older egg will lie almost on its side, with its broad end up. If an egg floats to the top, discard it.

Keeping bees

A swarm of bees in May is worth a load of hay;
a swarm of bees in June is worth a silver spoon;
a swarm of bees in July is not worth a fly.
Traditional saying

Keeping bees is part of the country tradition and old-time gardeners are likely to have had at least one hive. This would place you in exalted company, because their golden honey really is the food of the gods. As an infant on the island of Crete, no less a figure than Zeus – commander-in-chief of the Greek gods – was

sustained by nymphs who fed him honey. Further north, Odin – the one-eyed magician father god of the Norse pantheon – drank mead, made from honey, to acquire all-seeing wisdom.

There are thousands of different species of bee, found all over the world, except Antarctica. Social bees that live together in colonies include the western honeybee and the bumblebee, while solitary bees include the mason bee and leafcutter bee. Old country gardeners would have been most familiar with the honeybee – the inhabitant of their hives.

Honey cough balls

Honey naturally granulates, at different rates depending on where the nectar comes from, and certainly if it's allowed to get too cold. If you want to make it runny again, warm it slowly so as not to spoil it. Here is a lovely old-fashioned remedy using honey. For a ticklish cough, blend a dessertspoonful of granulated honey with a knob of butter, and form into little balls. Dissolve slowly in the mouth. Children especially will like this.

The secret life of bees

All bees are real grafters, but honeybees are especially wondrous. Here is some insider knowledge on these incredible insects:

- Making honey is really hard work: to produce sufficient to fill a 450 g (1 lb) jar, a honeybee has to gather nectar from about two million flowers.

- When honeybees find a good source of nectar, they do a sort of 'waggle' dance to pass on the good news to each other.

- We know that a dog's sense of smell is more powerful than ours – fifty times stronger, it is claimed. A honeybee can do even better: its sense of smell is said to be fifty times more powerful than that of a dog.

- Bees can see purple more clearly than any other colour, but they don't see red at all.

- Making honey isn't an instinctive process – young worker bees have to learn this skill from their older companions.

- Honey tastes different depending on the flowers from which bees have gathered the nectar: clover produces a light, delicate honey; heather produces a darker, thicker and more strongly flavoured honey.

Honey

Beeswax furniture polish

The honeyed scent of beeswax furniture polish makes a home feel so welcoming. Here is a simple, old-fashioned recipe: shred 28 g (1 oz) of beeswax into a jar and pour a teacup of turpentine over it. Place the jar in a bowl of warm water and leave in a warm place for the beeswax to dissolve slowly. Remove the jar from the water and put on the lid. All you need now to achieve glowing wood surfaces is a little polish and a lot of elbow grease.

Treating bee stings

If, unfortunately, you do get stung by a bee (or a wasp), here is some fifteenth-century advice:

*A plaster made of wilde malowe leaves
is good to draw out the stinge.
The donge [dung] of a goose draweth
out the venom of any wasps.
And salt and vinegar tempered with hony is very good.
Oyle of bay is good also for the stynge.*

Jacob Meydenbach, *Hortus Sanitatis* (The Garden of Health, 1491)

6

The Flower Garden

*Bread feeds the body, indeed, but
flowers feed also the soul.*

The Koran

For many people, the word 'garden' means only one thing: an outdoor space filled with flowers. The ornamental garden can certainly be a source of joy, delighting the senses with its colours, shapes, perfume and the hum of insect life. The flowers we most associate with traditional gardens are the old-fashioned ones, such as roses, of course, sweet peas, wallflowers, foxgloves, hollyhocks, violets, poppies, cornflowers, nasturtiums, larkspur and many others.

Getting the look

We can complain because rose bushes have thorns,
or rejoice because thorn bushes have roses.

Abraham Lincoln (1809-65), sixteenth US President

The typical old country garden would have been a delightful jumble of unpretentious, brightly coloured flowers that had been raised from seeds, cuttings or bulbs, or that had freely self-seeded themselves. As well as looking charmingly old-fashioned, this was a practical, working space where plants had to earn their keep; some of the flowers would have been grown for medicinal purposes alongside vegetables to provide food for the kitchen, and hens that were kept to provide eggs. The key to this style of garden is informality, and that's great news because it means you don't have to work too hard to achieve the look.

As well as knowing what to include in your planting, you also need to know what *not* to include. Here are some plants to avoid if you want that old-style country look:

- 'Architectural' specimens such as phormiums or palms, as these belong in a modern, urban garden.
- High-maintenance, tropical-style plants such as banana palms and cannas, which, in cooler climates, need wrapping in horticultural fleece or lifting and bringing in for the winter – too much fuss, an old-timer would say.
- Box topiary that typifies the gardens of grand houses (but the odd small clipped shape of, say, an animal, would add a touch of humour and not be out of character).

The following list will give you a few ideas to start with. The flowers have been divided according to their most useful characteristic – such as scented or shade-loving – but some may belong in more than one category; for example, sweet violet is listed under 'Scented favourites' but can also be grown in shade.

Don't forget vertical surfaces either. Let honeysuckle and jasmine clamber up walls and fences or, for that really classic look, grow roses around the door.

Winter and spring favourites

Anemone . bluebell . buttercup . cowslip . crocus . daffodil . forget-me-not . heather . hellebore . lilac . peony . snowdrop . tulip

Summer and autumn favourites

Alyssum · aster (Michaelmas daisy) · catmint · cornflower · French marigold · globe thistle · hollyhock · lady's mantle (alchemilla) · larkspur · love-in-a-mist · lupin · mallow · nasturtium · pansy · penstemon · petunia · phlox · poppy · pot marigold (calendula) · rose · scabious · sea holly · snapdragon · sunflower · yarrow

Globe thistle

Scented favourites

Dame's rocket / sweet rocket · evening primrose · jasmine · lily ·
lily of the valley · mock orange · night-scented stock · phlox ·
stock · sweet pea · sweet violet · sweet william · tobacco plant
(nicotiana) · wallflower

Lily of the valley

Favourites for shade

Aquilegia (columbine) . bugle . Canterbury bell . cranesbill (hardy
geranium) . false goatsbeard . foxglove . lungwort (pulmonaria) .
Solomon's seal (polygonatum) . wood spurge (euphorbia)

*It is a golden maxim to cultivate the garden for the
nose, and the eyes will take care of themselves.*

Robert Louis Stevenson (1850–94), Scottish novelist, poet and travel writer

Right plant, right place

Remember the classic gardener's mantra, 'right plant, right place'. If you know a plant's original natural habitat, you'll have a better chance of giving it the location and conditions it likes, and hence greater growing success. Plants that are woodland natives, for example, are happy in shade. Similarly, early spring bulbs, such as the bluebells that carpet woodland floors, will also tolerate a shady spot. They have evolved to flower in dappled shade before the trees are fully in leaf.

Back in time

Before the invention of more modern materials, sunflower stems were used to fill lifejackets and give them buoyancy, while dried nasturtium seeds were ground into powder during World War II for use as a pepper substitute.

Plant explorers

Growing native wild flowers is gaining in popularity, but in fact *all* the flowers we grow in our gardens were once wild, native plants somewhere in the world. It is due to the work of plant collectors and breeders that we have such a wealth of flowering plants to choose from today.

The nineteenth century was the great age of the plant hunter, who was generally male. Intrepid women travellers tended to paint the flowers they saw, rather than bring specimens home. Two notable female botanical artists were the Swiss naturalist Maria Sybylla Meriam (1647–1717), and the British Marianne North (1830–90).

Storms, pirates, bandits, angry mobs and landslides were just some of the hazards that these Victorian plant seekers had to endure, but what helped to ensure success was the invention of the Wardian case – what we would now call a terrarium. This enclosed glass dome protected the precious specimens they had collected on the long voyage back to Europe. The descendants of the wild forms these explorers discovered grow in many gardens today. The species name – *Rhododendron forrestii*, *Rhododendron fortunei*, for example – often holds a clue to who discovered what.

- More than two hundred different plant species, including azaleas, chrysanthemums and tree peonies, were introduced to Western gardens by Robert Fortune (1812–80) from China and Japan. Fortune's most famous mission to China, however, was as an industrial spy for the British East India Company, who tasked him with discovering the secrets of tea production, on which China then had a monopoly. In disguise and speaking Mandarin, Fortune was able to infiltrate a tea factory to observe the process for himself. He also enabled the shipment of around twenty thousand *Camellia sinensis* tea plants to India and thus helped to establish the tea industry in India.

- The regal lily was introduced to the West by Ernest 'Chinese'

Wilson (1876–1930), who found them growing in southwest China.

❧ A species of rhododendron and the parent of the modern camellia were found by George Forrest (1873–1932) in China.

❧ The wild sweet pea, ancestor of the modern variety, is thought to have been discovered by a Sicilian monk, Father Francis Cupani, around 1695. The Scottish horticulturalist Henry Eckford bred the modern sweet pea in the late 1800s.

❧ The nasturtium is descended from Peruvian species that Spanish conquistadors brought back to Europe in the late fifteenth century.

❧ The sunflower originates in the Americas, where it began to be grown as an important food crop by Native Americans around 4,500 years ago. Europeans came across it in the early sixteenth century, and brought it back to Europe. These are real sun-worshippers and turn their heads to follow the sun in a natural process called heliotropism.

Tulipomania

The tulip is said to have originated in Turkey, and takes its name from its shape, which resembles a turban: *duliband* in Turkish. In seventeenth-century Holland, people went wild for this new arrival from the

Tulip

East. Bulbs were traded for ridiculous amounts of money, and the obsession known as 'tulipomania' was born. In 1637, bulbs of one species of tulip were offered for sale at between 3,000 and 4,200 guilders, which was more than ten times the annual wage of a skilled craftsperson. In today's money, some bulbs fetched the equivalent of £300,000. The famous tulip fields of Holland are a reminder of those heady days.

Flower power

Few flowers have such an ancient medicinal pedigree as the calendula, or pot marigold. Back in the late 1500s, the famous English botanist John Gerard recommended a concoction of marigold flowers and leaves as a remedy for 'red and watery eies'. During the American Civil War in the 1860s, doctors used dried calendula petals to treat wounds. Calendula leaves are what an Old Wife might have applied to a nettle sting. For a homespun calendula lotion, boil some of the flowers in water for about twenty minutes, strain and leave to cool, then apply to the skin.

Way before she got to be an 'Old Wife', a young woman might have needed a little help in affairs of the heart. The solution was obvious: brew up a love potion containing calendula. This bright flower was credited with other unworldly powers, too. Back in the sixteenth century, a sip or two of calendula potion was said to reveal the presence of fairies to you, while calendula petals scattered under your bed would protect you from being robbed during the night.

Larkspur, the annual form of the delphinium, was equally handy. It was believed to protect against any nasties you didn't want coming your way, from lightning, scorpions and snakes to witches and ghosts. For more practical and prosaic Europeans and Native Americans, it could also be pressed into service as a blue dye.

Some flowers are poisonous, but used judiciously their toxic components can provide the basis for some beneficial drugs. Digoxin, derived from foxgloves, is used in the treatment of some heart conditions. A drug called galantamine, used in the treatment of Alzheimer's, has been developed from a substance found in daffodil bulbs and those of the *Galanthus* genus.

Floral oil

Try this gorgeous old-fashioned recipe for creating perfumed floral oil. Steep some cotton wool in olive oil, and place a layer in an earthenware jar. Cover with a layer of scented flowers, such as roses, wallflowers, jasmine, carnations. Repeat the layers of cotton wool and flowers until the jar is full. Cover tightly and leave to stand in the sun for a week. Turn out the contents, discard the flowers, press the oil from the cotton wool and transfer to a bottle. The oil will now be scented and the cotton wool can be used to perfume your chests of drawers.

Edible flowers

Many flowers have made it to the table as edible ingredients. Here are just a few:

- The nasturtium has a long history as a foodie's flower. The Incas used it as a salad ingredient and medicinal herb, and nasturtium flowers and the peppery leaves still make a pretty addition to salads. Pickled in spicy vinegar, the seeds are a frugal alternative to capers, which are the flower buds of the *Capparis* plant.

- Primrose flowers were once fried in butter and sugar and served as a dessert, a dish that was much appreciated by Benjamin Disraeli (1804–81), a British prime minister and favourite of Queen Victoria.

- Calendula, the 'pot marigold', was added to the pot to flavour stews and soups, to add a rich yellow colour to butter, cheese and custards, and as a cheaper alternative to saffron, which is derived from crocus flowers and is one of the world's most expensive spices.

- Rose petals and violets can be crystallized in a little beaten egg white and caster sugar, allowed to dry, then used to decorate cakes or desserts.

Put a plate of flowers of the nasturtium in a salad bowl, with
a tablespoonful of chopped chervil; sprinkle over with your
fingers half a teaspoonful of salt, two or three tablespoonfuls
of olive oil, and the juice of a lemon; turn the salad in a
bowl with a spoon and fork until well mixed, and serve.

Turabi Efendi, *The Turkish Cookery Book* (1864)

Rose petal jelly

Bring the fragrance of the garden to your table
with this beautiful old recipe for rose petal jelly.
You will need 2 breakfast cupfuls of rose petals, 2
breakfast cupfuls of warm water, 2 ½ cups of sugar, 2
tablespoons of runny honey, 1 teaspoon of lemon juice
and, if wished, a drop or two of red food colouring.

Slice the petals into strips about 1 cm (½ in.) wide,
discarding the bases. Simmer the petals in the water
for about 10 minutes, or until tender. Strain the liquid,
reserving the petals; the liquid should have reduced to
about 1 ½ cups. Combine it with the sugar and honey,
boil very gently for 5 minutes, then add the reserved
petals and simmer over a very low heat for 40
minutes, stirring it often and watching to see that it
doesn't burn. Add the lemon juice and red colouring,
if using, and simmer for a further 20 minutes.

Pour into hot, sterilized jars and seal at once.

Keep them coming

Flowering plants exist to produce seeds for the next generation. You can exploit this by cutting off dead or dying blooms to encourage new ones to form. It's standard gardener's practice and is known as deadheading. When you remove old flowers, cut back to a joint or all the way to the bottom of the stem, so that the stems don't die back.

Keep them fresh

Try these clever ways of prolonging the life of your cut flowers:

- Plants with hard stems, such as roses, will last longer if you split the stems and peel back the skin at the bottom. Some flower arrangers like to crush the stems of flowers before placing them in a vase; this is said to prevent the cut ends closing up so that the stems continue to absorb water.
- Place the stems of cut flowers in some very hot water for about two minutes, then transfer them to a vase of cold water containing a dash of salt.
- Wrap tulip stems in newspaper, place them in water up to their necks, and leave them to soak for a few hours before arranging.
- Adding sugar to the water for delphiniums, and charcoal to the water for daffodils and narcissi, is said to prolong their life.

- Never mix daffodils with other flowers as they secrete a poison that is toxic to other blooms.
- Add a dash of vinegar to the water before placing your flowers in the vase; this slightly acidifies the water and helps to slow the growth of bacteria. A copper coin in the water should have the same effect.
- Always remove any leaves that will be below the water line, or they will rot.
- If cut flowers are wilting, they need a hit of sugar. Cut the stems off at an angle about 2.5 cm (1 in.) from the end. Clean the vase, fill with lukewarm water and stir in three teaspoons of sugar. Place the flowers back in the vase. Leave them to have a good drink and they'll soon perk up again.

Rose or violet vinegar

Make some beautiful flower vinegar for your own use, or to give as a gift. For rose vinegar, fill a jam jar with rose petals, fill with white wine vinegar, cover tightly and stand in the sun or a warm place for three weeks, to infuse. Strain and bottle. For violet vinegar, half fill a jar with violets and top up with boiling white wine vinegar. Leave to get cold, then strain and bottle.

Say it with flowers

Flowers are words which even a babe can understand.

Bishop Arthur Cleveland Coxe (1818–96)

For centuries, certain symbolism has been attached to particular flowers in countries around the world. But it was perhaps the Ottomans – members of the Turkish dynasty who ruled the Ottoman Empire – who were first to adopt a floral language, using flowers and other items to send coded messages. The practice was described in a letter written in 1718 by Lady Mary Wortly Montagu, when she was living in Constantinople (now Istanbul) with her husband, the British ambassador to Turkey. The idea took hold in Europe and then America, but reached the peak of its popularity in the Victorian era. Flowers offered reticent Victorians a way of delivering messages that they did not dare to speak aloud. Families even had flower 'dictionaries' to help them translate this secret language.

To understand it fully you had to be extremely observant, because even where the ribbon was tied, or the way in which the flowers were presented, could be significant. Being a great traditionalist, the Old Wife would likely have been able to crack the code.

There isn't a single floral language and the symbolism of individual flowers vary from one source to another, but here are some common meanings that you might have found in a Victorian flower dictionary:

Anemone – loss, grief (because the anemone closes its petals at night, fairies are said to sleep inside)

Bluebell – kindness

Larkspur – levity

Lily of the valley – purity

Marigold – grief

Narcissus – boastfulness (the flower is named after a beautiful youth in ancient Greek myth, who spent all his time gazing at his reflection)

Nasturtium – patriotism

Poppy – consolation

Red rose – romantic love

Red tulip – declaration of passionate love

Sweet william – gallantry

Wallflower – faithfulness

7

The Wildlife Garden

*You can drive out Nature with a pitchfork,
but she keeps on coming back.*

Traditional saying

Picture the scene: it's a warm summer's afternoon and you are lazing in a hammock, lulled by the drowsy hum of bees and the song of birds. Is this Heaven? No, it's your own garden. Back in 1888, William Butler Yeats conjured just such a scene in his famous poem *The Lake Isle of Innisfree*. He described how he would plant nine bean-rows, have 'a hive for the honey bee' and 'live alone in the bee-loud glade'.

Despite widespread species-decline since the days of nineteenth-century poets, it is not impossible to create your own little 'bee-loud glade' by developing the kind of environment that is welcoming to wildlife. Gardeners can be important eco-warriors, doing their bit to sustain wildlife and the environment.

The really good news is that if you encourage Nature, she will help you in return and this means:

- No chemicals – no pesticides, herbicides or fungicides
- More bees, butterflies, birds and other wildlife to bring a sense of wonder into your garden
- Less toil and more lazing about – Nature is working on your behalf while you do so
- Smug superiority – you know you're doing the right thing

The aim is to see your garden as a self-sustaining eco-system, as much as possible. That means adjusting the balance: inviting natural predators, pollinators and processes to take over and do a lot of the work for you. It may take time to get the balance right, but have faith and persevere. Establish conditions that birds, insects and other wildlife will like, and they will – uncannily – find their way to you. In the old days, this kind of gardening didn't have a label. It wasn't called 'organic'; it was just what gardeners did.

Enticing pollinators

Follow the old tradition and tempt natural pollinators into your garden by growing their favourite, nectar-rich flowering plants. As a general rule of thumb, it's not the big, blowsy showstoppers they go for, such as roses or dahlias: the petals just get in their way. What they prefer are simple, open flowers that allow them easy access to the nectar at the base of the petals. They also seem to like tiny flowers that, to us, look insignificant. As they drink the nectar, these essential workers pick up pollen which they then carry to the next flower of the same species and – hey presto! – pollination occurs. Seeds develop and the next generation of that plant is assured, in a perfect symbiotic partnership.

Buddleia

As well as the promise of nectar, flowers have other ways of luring pollinators. Some do it with scent and some with their shapes and bright colours; others use markings on their petals known as 'nectar guides' to lead pollinators in the right direction.

These bold entry signs are only visible to creatures with ultraviolet vision, such as bees – humans cannot see them at all. Scientists have also discovered that flowers that sway on long, thin stems are better at drawing the attention of pollinating insects – it's as if they are calling out 'over here!'

Old-time gardeners may not have known all the science behind such floral seductions, but they will have been aware of the value of pollinating insects in their gardens.

Plants for pollinators

Try to grow plants that will flower in different seasons so that you can feed pollinators throughout the year (except for winter, when they are largely dormant).

If bees had a Top Ten chart, lavender would be at Number One. Anyone who has grown a lavender bush will know that it's always humming with bees of every kind, drawn by the siren call of its scent and sweet nectar. For butterflies, it's buddleia. In fact, such a magnet is it to these insects that it has earned itself the nickname 'butterfly bush'. In autumn, the honeyed scent of the humble flowering ivy is irresistible to late-flying bees and other pollinators who come to drink their fill before the winter sets in. The blackberries that follow provide a feast for blackbirds.

These days, seed packets and plants sold in garden centres often have a 'bee-friendly' label that can help you make the best choices. What could be more joyful than welcoming bees and other pollinators into your garden? The following are some suggestions for flowering plants that can make this happen.

Late winter / spring

Bugle
Crocus
Forget-me-not
Fruit-tree blossom
Heather
Lungwort (pulmonaria) – this has tiny, pink and blue
   flowers that bees love
Mahonia
Rosemary
Single-petalled hellebore
Snowdrop
Willow catkins
Winter aconite
Winter-flowering clematis

Summer

Allium
Aquilegia (columbine)
Bergamot
Blackberry (late spring /
early summer)
Borage
Coneflower (echinacea)
Dog rose
Evening primrose
Globe thistle
Hardy geranium

Borage

Poppy
Scabious
Sea holly (eryngium)
Sunflower

Late summer / autumn

Aster (Michaelmas daisy)
Cornflower
Ivy
Japanese anemone
Salvia
Sedum
Teasel
Verbena

For long-tongued bees

The nectar of tubular-shaped flowers is accessible only if you have the right equipment. Bees such as the common bumblebee have the requisite long tongue, but they still have to get on in there to access the food – just watch the wiggling, furry bottom of a bumblebee disappearing deep into a flower cone! Flowers such as these will feed these furry friends:

- Foxglove
- Honeysuckle
- Penstemon
- Snapdragon

For moths

You may not realize it because they are active only at night, but moths are important pollinators, too. Bright colours and patterns are of little use in low levels of light, so flowers have had to evolve another way of tempting in these helpers: perfume. Grow evening-scented plants and you too will benefit, as you sit outside on a warm evening surrounded by wafting floral fragrances. Here are a few to try:

- Evening primrose
- Honeysuckle
- Jasmine
- Nicotiana (tobacco plant)

Back in time

The common names that people once gave plants often provide a clue to how they were traditionally used. For example, pulmonaria, or lungwort, has spotted leaves which herbalists once thought resembled a lung, so the plant was used to treat lung disorders. The prickly dry heads of teasel were once used by textile workers for combing fabric in the textile industry – hence the word 'tease'.

Nasty nettles?

All nettles do is sting and they are good for nothing else, right? Think again. In fact, nettles play an important part in the natural food chain. It's another example of Mother Nature's brilliant symbioses. Here's how it works:

1. The hollow hairs on the leaves contain a venom. When you brush against the fine hairs, they break, releasing their stinging poison. This is the nettle's way of protecting itself from any creatures that might want to eat it, such as grazing animals.
2. The stinging nettle leaves give shelter to numerous species of small invertebrate, such as ladybirds, allowing them to thrive. Some butterfly and moth species, such as the red admiral, small tortoiseshell and peacock butterflies, depend heavily on nettles for food and as breeding sites.
3. The increased invertebrate life brings in natural predators such as frogs, hedgehogs and birds, to feast on the bounty that the nettles have generated.

You won't want to give nettles free rein in your garden, but perhaps you could allow them a corner in which to flourish, and help Nature at the same time. If you do need to remove some, remember to wear thick gardening gloves and cover your arms. Once you have severed the stems from the roots, the poison will stop flowing and the leaves will no longer sting. You could even do a spot of wild foraging and turn young nettle leaves into a nutritious soup, or try your hand at making a recipe that's said to date back thousands of years: nettle pudding – a mix of nettles, barley flour and water.

... there we did eat some nettle porrige, which was made on purpose to-day for some of their coming, and was very good.

Samuel Pepys, diarist, 25 February 1661

Old-timer's tip

If you went for a walk with a wise old gardener and you accidentally got stung by nettles, the first thing they would do is grab the nearest dock leaf and rub it on the sore patch. This old folk remedy is thought to work because rubbing the leaf releases its cooling, moist sap. Back home, your friend might dab the affected area with a clean cloth and vinegar.

Wild or weed?

A weed is but an unloved flower.

Ella Wheeler Wilcox (1850–1919), US writer and poet

Not that long ago, gardeners were being encouraged to zap weeds with heavy-duty chemical herbicides (some of which are now banned). Nature had to be shown who was boss! The trouble is that this approach resulted in highly manicured, regimentally controlled gardens that were not particularly wildlife-friendly. What were classified as 'weeds' were in fact some of the very plants that helped to sustain wildlife.

What is a 'weed' anyway? Some maintain that it's just a flower in the wrong place. And you have to give these wild plants – for that is what they are – credit for being so robust and reproducing so successfully. The problem with weeds is that they can be just a bit too successful and can compete with more tender specimens for nutrients, water and space. One example is the highly prolific dandelion. One flower can produce about 180 seeds, but leave it to grow for three years and it may generate as many as 5,000!

Another equally determined colonizer is bindweed. It does have pretty white flowers, it is true, but unfortunately it's a bit of a thug. Like Jack's beanstalk shooting up to the sky, it will wind itself around any support it finds and smother the cultivated plants that you are trying to grow. Bindweed is the ultimate survivor – the tiniest bit of root left in the soil will regrow, and gardeners once felt they had no option but to resort to some pretty nasty chemicals to eradicate it.

So just how do you achieve that balance between leaving some wildlife-friendly weeds while at the same time not allowing them to take over your garden? It's all a matter of being a bit more relaxed and using non-toxic methods of eradication where necessary.

When weeding, the best way to make sure you are removing
a weed and not a valuable plant is to pull on it. If it
comes out of the ground easily, it is a valuable plant.

Anonymous

Prevent germination

If you're dealing with annual weeds such as chickweed, the trick is not to allow them to set seed. If the seed germinates, you'll have a whole new generation of weeds. You can hoe them, but do this before flowers have formed. Another solution is to refrain from digging the soil, as this exposes the seeds to light, which encourages germination.

One year's seeding makes seven years' weeding.

Traditional saying

Block out sunlight

Even tough perennial weeds, such as bindweed and couch grass, need sunlight on their leaves to photosynthesize and grow. (Photosynthesis is a process whereby plants convert sunlight into sugar and energy.) Deprive them of this vital source and they will eventually give up. Cut the tops off, cover with a few layers of cardboard, then top with a 10–15 cm (4–5 in.) layer of compost. Or you can use proprietary weed-suppressing sheeting.

The water cure

Don't waste your time trying to dig out weeds between paving slabs. Here's a trick that is so simple and effective it's hard to believe – just pour boiling water on them! This might seem a bit cruel, but isn't it kinder than using some nasty chemical? The weeds will die back down to the roots, though they might smell a bit like boiled cabbage during the process. Another clever trick is to sprinkle them with salt, then wait for the rain to do the work.

Bring in the competition

Grow Mexican marigolds among the weeds. Their roots produce chemicals that can check the growth of neighbouring plants – even such tough ones as bindweed. They'll have the same effect, though, on more desirable plants, so use this method only on a weedy patch that you want to clear.

> *Cut a thistle in May, it will be back the next*
> *day; cut a thistle in June, it will be back soon;*
> *cut a thistle in July, it will surely die.*
>
> Traditional saying

Hoe, hoe, hoe

The hoe is one of the most useful tools in the gardener's armoury. Keep hoeing and even tough perennial weeds such as bindweed will give up. Of course, it's only possible to hoe if there is sufficient space between plants, or you will risk slicing through them, too. Also, take care not to hoe too deeply, as you may bury weed seeds, which will then germinate. It's more effective to hoe in dry weather.

The helpful dandelion

The prolific dandelion gets its name from the French *dents de lion*, or 'lion's teeth', after the shape of the leaves. It has a French nickname, too – *pis-en-lit*, or 'piss-in-bed' – because of the diuretic effect of its leaves, which are often eaten in salads. The ground-up roots are traditionally used to treat heartburn or as a laxative, while the flowers can be made into wine or tea.

Enticing predators

Companion planting, described in the Pests and Diseases chapter, forms part of natural, old-style garden management. Companion plants deter pests but they do more, too: they actively attract predators that eat the pests that eat your plants.

Greenfly, blackfly and other aphids are a good example of how this works. These tiny insects suck the sap of plants, and can distort their growth and spread plant diseases. A clue to their presence is a trail of ants going up and down stems. Ants and aphids have a symbiotic relationship. Ants 'milk' the honeydew, a sticky sugary liquid that aphids secrete, and in return the ants give the aphids some protection against predators. Grow certain flowering plants to bring in hoverflies, lacewings and ladybirds and they will feast on the aphid population. Many of these natural predators will pollinate your plants, too.

- Sage, fennel and dill flowers attract hoverflies.
- Calendula and wormwood flowers entice ladybirds, lacewings and hoverflies.
- Hop, a decorative climber, attracts ladybirds.
- Nasturtium flowers are popular with aphid-eating insects.
- Borage, with its delicate blue flowers, attracts hoverflies, as well as bees and butterflies.

Other garden friends

Here are some other helpful allies, who will gobble up slugs, snails and the odd unwanted caterpillar:

- Hedgehogs
- Frogs and toads
- Ground beetles

Convince them that your plot is the place to be by being a little untidy. A pile of leaves or logs left in a quiet corner will provide them with the ideal hidey-hole.

A word of warning: avoid using slug pellets (especially the

ones containing metaldehyde) which, as well as killing slugs and snails, can be toxic to other wildlife. Try other methods of controlling these pests (see the chapter on Pests and Diseases) and aim for a manageable balance of predator and prey, rather than total elimination.

Hedgehog

Hedgehogs

These prickly mammals that live in Britain, Europe, Asia, Africa and New Zealand are only out and about at night, so they may secretly visit your garden – even if you live in a city – without your knowing it. Here are some ideas for how you can help:

1. Don't fence hedgehogs in! Walls and fences are unnatural barriers that prevent animals from foraging, looking for mates and breeding. Given the choice, a hedgehog can wander over a total of about 1.5 km (1 mile) in an evening in search of food. Allow hedgehogs room to roam – dig holes under fences and, if possible, make openings in walls.

They need only be 12–15 cm (5–6 in.) wide. Best of all, get together with your neighbours to create a series of openings across all your gardens to form a hedgehog super-highway.

2. Provide a safe place for hedgehogs to nest and hibernate. They appreciate piles of leaves, twigs and logs, and even your compost heap. Hedgehog delicacies such as slugs and beetles will take up residence there, too, so you'll also be providing them with a natural source of food.

3. Put out food to supplement a hedgehog's diet only in winter or dry weather, when juicy slugs and other treats are harder to find – you don't want it getting too reliant on you. Offer plain, meat-based cat or dog food, or special hedgehog food but never bread and milk, which can dehydrate and even kill hedgehogs. Vary where you set out the meals to mimic what would happen in the wild. If necessary, tuck the food in a place cats cannot reach. Put it out at dusk and, if it doesn't get eaten, remove it and replace it with fresh food.

4. Also leave out a shallow dish of water.

Bring in the birds

The north wind doth blow, And we shall have snow,
And what will the robin do then? Poor thing.

Traditional nursery rhyme

Watching and listening to birds in the garden is one of life's great pleasures. These songsters are beneficial, too, since ground-feeders, such as blackbirds and thrushes, will feast on pests. They don't ask for much in return: just safe shelter, food and water.

1. Hedges provide a great place for birds to nest. Avoid cutting them during the breeding season from spring through to summer.

2. Support feathered housebuilders by growing, or making available, the plants and materials that they use to construct their nests. Blackbirds, for example, like dry grasses and plant stems, bulked up and held together with moss and mud. Tiny wrens like plenty of moss, with leaves and dry grasses. Having these building supplies close by means birds won't need to travel further afield to find them, thus saving them time and energy, and making it more likely they will nest in your vicinity.

3. Grow plants that will provide a source of natural food. The densely seeded heads of sunflowers and teasels are popular items on the menu.

4. Set out feeders containing seed mixes. It's more cost-effective to buy better-quality mixes since the cheaper brands are bulked out with millet and wheat which seem to appeal only to pigeons.

5. Peanuts are a great, high-energy food, but buy yours from a reputable supplier who will have tested them for afloxin, a carcinogenic mould that can be fatal for birds. Store the nuts in a dry, sealed container – not in the plastic bag they arrived in – to keep them dry and prevent moisture and mould. Outside, peanuts should keep for about a week, but replace them with fresh ones if they start to look discoloured or mouldy.

Robin

6. Never scatter peanuts on the ground as smaller birds can choke on them. Offer them in a metal mesh peanut feeder.

7. Avoid hanging feeders made out of plastic mesh, as birds' legs can get trapped in them.

8. It's now advised to feed birds all year, not just in the winter. However, they won't need to dine in your garden so often in the warmer months when natural foods are in greater supply, so don't overfeed then. If you notice that food is going untouched – and potentially going mouldy – discard it and replace with smaller amounts.

9. Birds need to feel safe from predators such as cats and sparrowhawks, so place feeders in sheltered spots where birds can easily escape to greater cover.

10. Minimize infection by bacteria and fungi by cleaning bird feeders and tables regularly, perhaps as much as weekly. Discard old food, rinse the feeder and give it a good scrub with soapy water and a weak disinfectant solution, rinse again, and leave to dry.

11. Move bird feeders regularly, say once a month, to avoid the build-up of bacteria in any one spot.

12. As well as food, birds need water: not just to drink, but for bathing in to help keep their feathers waterproof. A bird bath or any other suitable container will do. But remember to site it out of easy reach of potential predators, and apply the same levels of hygiene as you do for the feeders. Placing a stone or two in the container will give smaller birds something to perch on.

> *You cannot catch old birds with chaff.*
> Traditional saying

Don't forget water

Creating a pond in your garden is one of the best things you can do for wildlife. Birds, hedgehogs and even bees need the odd drink of water, and frogs and toads are dependent on it for breeding. Create a pond in your garden and you'll be amazed at how quickly the news travels – before you know it, amphibians will appear from nowhere to populate it. Make it safe for non-aquatic wildlife by giving it sloping sides or some other means of exit, such as stacked bricks, piles of stones or a log 'gangway', so they can clamber out, and so that bees have somewhere to perch while they drink.

Grow plants such as water iris to give dragonfly and damselfly larvae a climbing frame to ascend when they are ready to turn into adults. Remember that at least half the surface should be covered with plants to shade the water and prevent algae taking over. Pond plants will also provide thirsty insects with a platform to drink from without danger of drowning.

If a thick layer of ice forms over the surface in winter, this effectively seals the pond off and oxygen levels in the water may drop so low that the inhabitants (frogs and fish) cannot breathe. You may be tempted to smash the ice – *don't!* The impact could send dangerous, and potentially fatal, shock waves into the water. Instead, create a 'breathing hole' by floating a plastic ball on the surface which you can lift off during the day, or very gently thaw the ice with warm water to let in some oxygen.

Back in time

In medieval Europe, fishponds were common features in the grounds of monasteries and grand houses. These were not created for the benefit of wildlife or to please the eye, but to provide a reliable source of food. Carp and pike were two species that were popularly farmed.

Danger – keep out!

There are risks to wildlife in the most unexpected places in a garden. Hedgehogs may be nesting in your compost heap or bonfire pile, and frogs may be taking shelter under plants or bushes. Before you start that fire or strim long grass, check to see if any animals are hiding there, and don't fork over your compost heap.

8

Pests and Diseases

A garden is a grand teacher. It teaches patience and careful watchfulness; it teaches industry and thrift; above all it teaches entire trust.

Gertrude Jekyll (1843–1932), British horticulturalist,
garden designer and writer

Sucking, slicing, gnawing, biting, burrowing – garden pests of every stripe are feasting on your plants. Meanwhile, rusty spots, black patches and a powdery white dust are appearing on leaves. Stone fruits, speckled with white dots, are going brown and shrivelling up like wizened prunes. Just what is going on here? And what old-fashioned wisdom can you draw on to deal with these problems?

Friend or foe?

This easy, rule-of-thumb method can help you identify which creatures are intent on eating your plants, and which are intent on eating each other. If they run fast, they're predators who need that speed to catch their prey. If they're sluggish, they're vegetarians who just want to loll about munching on greenery and flowers.

Old-timer's tip

An old way of dealing with carrot root fly is to mix crumbled mothballs into the soil where you are growing carrots. For cabbage root fly, wrap a strip of kitchen foil around the roots of young cabbages. Another old remedy against both pests is to lay a length of creosoted string between the plants.

Helpful neighbours

Canny old gardeners have long known that growing certain plants in among others they wish to preserve can help to deter pests. This practice is known as 'companion planting'. Some strongly aromatic companion plants – which include many herbs – work by masking the scents that attract pests to their neighbours. Other companion plants work as decoys or 'sacrificial victims', being so irresistible to bugs that they choose to feed on them instead.

Many companion plants have the added benefit of drawing in natural predators, to create a more balanced ecosystem in your garden. For suggestions on which plants to grow in order to entice these allies, consult The Wildlife Garden chapter.

Borage If you live in North America or Australia, you might be plagued by the tomato hornworm, which is the larva of the five-spotted hawkmoth. Grow this herb next to your tomatoes, which are this pest's favourite food. They also like other plants in the *Solanaceae* family, such as aubergines (eggplants) and peppers. Borage is said to improve the flavour of strawberries if grown nearby.

Calendula Grow these bright orange flowers – the traditional pot marigold of English cottage gardens – next to tomatoes to deter whitefly, and to tempt aphids away from beans.

Fennel If you allow fennel to flower, the blooms will attract hoverflies, which eat aphids.

French marigolds Grow next to tomato plants to keep aphids away.

Garlic Plant a clove next to a rose bush in order to deter aphids. The rose roots absorb chemicals from the garlic that aphids don't like, so this is effectively a systemic insecticide. But don't worry, you won't end up with a garlic-scented rose. Other members of the garlic family work too, but not as well as garlic.

Garlic chives Plant next to carrots to confuse carrot root fly. The adult flies can sniff out carrots from up to a mile away, and their larvae will drill through the roots.

Lavender Place next to carrots and leeks to fool aphids.

Mint Cultivate this highly aromatic herb to mask the irresistible scent of tomatoes, carrots, onions and brassicas, and to deter flea beetles. But remember: mint likes to run rampant, so grow it in a pot nearby to stop it taking over.

Nasturtiums Use as a 'sacrificial crop'. Blackfly are particularly fond of them. They will lure aphids away from your French and runner beans, and can tempt caterpillars away from your cabbages, too.

Rhubarb Add chopped leaves to planting holes before adding cabbage seedlings. This will help to deter cabbage root fly.

Sage Place next to plants of the cabbage family. Its strongly aromatic leaves will mask the scent of these brassicas.

Thyme Grow around your rose bushes to fool blackfly.

Wormwood Cultivate this unusual, strongly scented herb to ward off aphids and flea beetles.

Sage

Kitchen cupboard pesticides

Look on the shelves of garden centres and you'll still see a whole array of chemical pesticides. During World War II, there was a huge need to provide food for people, and land and Nature were subjugated to this end. Chemical sprays were used on farmland and the use of chemicals became the norm in the domestic garden, too. This strategy poses two problems: the first is that these products kill indiscriminately, destroying beneficial wildlife as well as unwanted pests; the second is that they can be a risk to human health. Luckily, many chemical pesticides that were once in general use have now been banned.

In the old days, of course, gardeners had to rely on traditional knowledge and what they could concoct themselves from natural ingredients. You might be surprised to learn that many of the key ingredients in these homemade pesticides are already in your kitchen cupboard.

Garlic spray

Garlic is a wonder ingredient with several uses in the garden. Its pungent smell is due to sulphur compounds it releases when the cloves are crushed: part of the plant's natural defence mechanism against anything that might want to eat it.

To make garlic water, roughly chop two whole cloves of garlic (you don't need to peel them). Place in a 1 litre (35 fl oz) jar, top up with water, cover the jar and leave to steep overnight. Then strain into a bowl through muslin or a fine sieve and stir in

1 tablespoon of soft soap. Pour into a bottle with a lid.

Garlic water is most effective when it's fresh. Choose a dry day and spray both sides of the leaves until soaking wet, using a solution of 250 ml (9 fl oz) of garlic water to 1 litre (35 fl oz) of water. Repeat weekly, or after heavy rain.

Works on: aphids, whitefly, slugs

Chilli spray

More adventurous old gardeners may have resorted to this remedy if they grew chillies in their greenhouses, or lived somewhere warm enough to grow them outside. Capsaicin is the ingredient that gives these peppers their fiery kick, and it delivers the same punch to garden pests as it does to humans. It also reduces pest populations by coating the eggs with an oily mix so they cannot hatch, and is a great deterrent against that most determined of garden nuisances, the squirrel.

Make it in the same way as garlic water (above), using a generous handful of peppers. Remember to wear gloves when chopping and don't touch your face. Steep the solution for two weeks before using. You can store it in the fridge for several weeks.

Use chilli water neat without diluting. Spray on the plants before insect infestations start, but don't apply it to leafy vegetables as they absorb the taste.

Works on: aphids, spider mites, whitefly, squirrels, rabbits, mice, deer

Tomato spray

Tomato leaves contain toxins that deter, and can even kill, certain sap-suckers. You'll need 500 g (1 lb) of leaves to 1 litre (35 fl oz) of water. Chop the leaves, soak them overnight in the water, then strain the mixture and spray on your plants.

Works on: aphids, red spider mite

Rhubarb spray

Like tomatoes, the large leaves of rhubarb plants are toxic. The leaf-to-water ratio is the same as for tomato spray (above), but you need to boil the leaves for half an hour before straining and spraying.

Works on: aphids, red spider mite, whitefly
Warning: rhubarb leaves are toxic to beneficial insects and to humans, too, so use with care and apply only to ornamental, not edible, plants.

Thyme spray

Soak a handful of thyme leaves in water, then strain and spray the infusion onto your cabbages.

Works on: whitefly

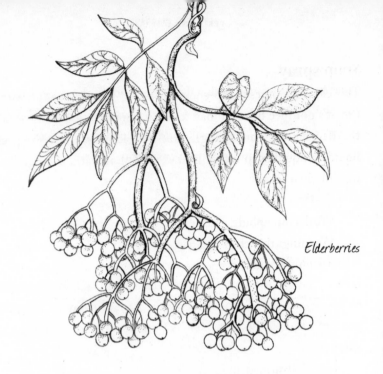

Elderberries

Mint or elder spray

Make an infusion with a mixed handful of spearmint, peppermint or elder leaves, strain, then spray the infusion onto or pour around vulnerable plants.

Works on: slugs

Elder

In the old days, gardeners would place a few elder branches under their gooseberry bushes to deter magpie moth caterpillars. These voracious eaters and others, including even rabbits, detest the smell.

Soap spray

This clogs the pores of sap-sucking pests so that they can't breathe (no, it's not nice but you may feel less sympathetic when you see the damage they are doing). Mix a tablespoon of washing-up liquid or liquid soap into 1 litre (35 fl oz) of warm water. Allow to cool, then spray.

Works on: aphids, caterpillars, whitefly, scale insects

Warning: this clogs the pores of beneficial insects too, so use only if you're desperate.

Old-timer's tip

A warm, damp greenhouse or cosy cold frame can provide ideal conditions for insect pests to proliferate. If an old-timer happened to have a cherry or common laurel tree in their garden or nearby, they might leave a bowlful of the crushed leaves in the greenhouse or cold frame overnight. The leaves, when crushed, give off a vapour containing prussic acid that is toxic to small insects. (If you do use this trick, you will of course take care, washing your hands after touching the leaves, and keeping them away from cats, dogs and horses.)

Slugs and snails

These pests can devastate a garden to such an extent that they deserve a section all to themselves. Collectively they are known as 'gastropods', which literally means 'stomach foot' and this is just what they are – sliding, slithering, oozing stomachs, moving about on a 'foot'. They feed using a kind of rasping tongue. It is sometimes even possible to hear them if you listen carefully at night, as they scrape away at their chosen meal. Popular gourmet treats for gastropods include young sunflowers, lobelias (they can strip a lobelia plant overnight) and, of course, their all-time favourite: hostas.

As with all pests, however, a degree of tolerance and compromise will make for a more relaxed gardener. It may be that you simply don't grow the plants that they especially love or, if you do, take extra steps to protect them. A trick with sunflowers is to start them off in pots and place them up high, where it is harder for these slurpers to reach them: for example, on a garden table. Then, when the young plants are big enough to withstand attack, you can transfer them to the ground.

Barriers

Slugs and snails slide along on a sticky slime trail, as the deposits they leave on shredded leaves will testify. One time-honoured method of defence is to put down a thick mulch of some rough or sharp material. The theory is that this will either dry out the slime on which the gastropod moves, or irritate it so that it won't cross the barrier. Mulches have mixed success but they are better than some other, crueller methods (such as sprinkling a slug with salt) – and they do help to retain moisture in the soil, as well. Experiment with the different options and see which work for you:

- Wood chippings
- Wool or hair (wool is now available in pellet form)
- Broken egg shells
- Pebbles
- Coffee grounds
- Straw
- Pine needles
- Cotton wool wrapped around individual stems
- Lengths of horsehair wrapped around the trunks of fruit trees (an old and perhaps somewhat eccentric trick)
- A greasy band of Vaseline smeared around the rim of plant pots (the slime trail of gastropods won't adhere so easily to this slippery surface)
- A band of copper wrapped around plant pots (the chemical reaction between the copper and the gastropod's mucus should stop the creature in its tracks)
- Copper coins laid on the soil, for a similar effect to copper bands wrapped around plant pots

⬛ Plastic drinks bottles, with the bottoms cut off and the screw-tops removed to allow air circulation, used as protective cloches for young plants (a commendable example of frugal recycling)

Traps

There are a couple of homemade traps you can make to snare your enemies. If they are still alive, see 'Do it by hand' (opposite) for how to dispose of them once you have caught them.

1. Place hollowed-out orange or grapefruit halves upside-down on the soil; these will provide tempting hiding places for slugs, which you can then dispose of the following morning.
2. Sink large, empty yogurt (or other) plastic pots halfway down in the soil, then fill them with milk and water, or with beer (the classic 'slug pub'). Drawn by the delicious scent, slugs and snails will climb into the pots and drown. Make sure, though, that the top of the pot is at least 2 cm (¾ in.) above the soil to prevent ground beetles climbing in, too – you'll want to save those, as they eat slugs and snails.
3. Create a cool, damp hiding place by laying a piece of damp cardboard or old carpet over the soil in some quiet corner, held in place with a stone. Gastropods will congregate under here, allowing you to catch them easily.

Pellets

If all else fails and you are desperate, you can resort to slug pellets. Although these aren't a traditional remedy, pellets have undergone a change for the better. Those based on ferric phosphate are a lot less toxic than earlier forms, but they must be used carefully in accordance with the manufacturer's instructions to ensure no damage to other wildlife, such as earthworms.

Do it by hand

An overlooked – but obvious – trick is simply to remove pests by hand.

- Aphids can simply be rubbed off with your fingers.
- Lily beetles and vine weevils can be picked off and crushed (Nature has made lily beetles conveniently visible, as they are bright red).
- Earwigs, which love to eat flowers under the cover of darkness, can be lured into a trap. Stuff a flowerpot with straw and up-end it on top of a bamboo cane; the earwigs will take up hiding there by day and can easily be disposed of.
- Slugs and snails can be caught in the act by torchlight at night, and moved. Lobbing them into your neighbour's garden is unlikely to be far enough, however. Snails have been shown to have a homing instinct and can travel up to 22 m (72 ft) a day, so throw them into some undergrowth at least that distance away.

Protecting edibles

The edible garden is prey to some very specific pests – ripe cherries are irresistible to birds, for example, and cabbages are gourmet fare for pigeons – and there are some practical and traditional methods applicable just to these crops:

1. Making a fruit cage or covering crops with netting to deter birds is a familiar protective measure, but make sure you use small-mesh netting, so that birds and small mammals do not get trapped. (The old practice of winding cotton thread among the fruit also risks birds getting trapped, so is best avoided.)

2. Burning your garden refuse under the branches of fruit trees is an age-old - if somewhat dangerous - method of deterring insect pests.

3. A trick that would not have been available to gardeners of yore is to hang old CDs – or anything sparkly – in the branches of fruit trees. The theory is that they will catch the light and flash as they move in the breeze, scaring birds away.

4. A mulch of straw is the traditional way of keeping strawberries off the soil – hence the name – and of deterring slugs and snails.

Old-timer's tip

Entice wasps away from your apples, pears or plums by offering them something even more delicious. Half fill a small jar with a mixture of beer and sugar or fruit slices, and cover the top with greaseproof paper. Make a small hole, about 1 cm (½ in.) wide, in the centre of the paper, then hang the jar from the branches. Lured by the irresistible scent, wasps will clamber into the jar but be unable to get out.

How to make a scarecrow

The idea of standing a human effigy in a field to scare away crows and other birds goes back to Ancient Egypt and Greece, but these early scarecrows were often a lot more ghoulish than their ragamuffin descendants. Evoking the form of a crucifixion and with dark cultural associations, some resembled witches or had heads made of rotting gourds (like a modern pumpkin-head) or animal skulls. In British regional dialect, a scarecrow is a 'hodmedod' or 'tattie bogle'; in Germany, simply a *feldpausch* (field bundle).

The crows that these figures are meant to scare away are far too intelligent to be taken in for long, but it's still fun to make a scarecrow. The simplest type uses two bamboo canes tied together to make a T-shaped frame, or perhaps two lengths of wood screwed together. The frame is then dressed in old clothes, and the head can be an old cushion cover stuffed with straw. For a fuller-figured scarecrow, stuffed tights make excellent head and arms: with the waistband tied, the stuffed top of the tights becomes the head, while the stuffed legs are the arms. A hat, 'hair' and facial features can be added.

So go on, get creative, but don't hog all the fun for yourself – share it with children, who will love this crafty activity.

Pets and other suspects

We love them but, let's face it, our pets can be a bit of a pain in the garden. Luckily, there are ways you can care for both your garden and your companion animal.

Cats

A patch of freshly dug, fine soil is like a magnet for moggies: clearly, this soft, pristine earth has been prepared solely for their use as a toilet. Whether the villain is your own cat or the feline from next door, there's nothing more disheartening than seeing your new seedbed dug up and defiled.

There are a number of canny, homespun tricks you can use against these toilet terrorists. They work in three ways: by smelling unpleasant – to cats, that is (these animals have a much stronger sense of smell than humans do, and dislike scents that we often find appealing); by creating a physical barrier; and by scaring the daylights out of the trespassers. You'll already have a lot of the basic ingredients at home, so these remedies may not cost you a penny.

- Scatter chopped-up citrus peel over the area you want to protect. Pungent citronella or eucalyptus oil will have a similar effect.
- Dust the area with chilli powder (this works for squirrels, too). You'll need to reapply after rain.
- Sprinkle crumbled mothballs on the area.
- Place teabags soaked in disinfectant around the site.

- Cover the area with a criss-cross grid of prickly or thorny twigs to form a barrier.
- Lay a length of hosepipe on the area; cats will mistake it for a snake and steer clear – until they realize their error, that is.

Dogs

Canines don't have the same toilet habits as their feline counterparts, but they may engage in a spot of manic digging in your flowerbeds. The reasons for this are complex and include instinctive behaviour, boredom, or a desire to escape into a cool hollow in hot weather.

If your dog does this, you'll need to explore the underlying reasons and work out ways to ensure that his or her needs are being met. While cats seem happy to laze around all day, dogs need more stimulation and more exercise, as well as sufficient play and human interaction. You could allow your dog a special digging spot in the garden or, if it's a persistent problem, you may have to cover a canine's favoured excavation site with stones, plant pots or some other obstacle.

There is one toilet habit that dogs have that can cause problems, and that's to do with urine. If a dog urinates on your lawn, it can scorch the grass and lead to unsightly dead, brown patches. It's worse with female dogs, who tend to squat to pee, while male dogs cock their legs up against vertical surfaces. The damage happens because the urine is high in nitrogen, which ironically is one of the ingredients in lawn fertilizer. The right level of nitrogen gives you lovely green grass, but too high a concentration burns it. The affected area of lawn will eventually recover, but you can help by acting quickly. Water the grass thoroughly as soon as you can to dilute the urine, then reseed later.

Mice and rats

An old mouse was running in and out over the stone doorstep, carrying peas and beans to her family in the wood.

Beatrix Potter, *The Tale of Peter Rabbit* (1902)

In traditional rural settings, these stealthy rodents might conjure visions of barns and farmyards, where mice and rats scurry about and raid the farmer's feed stores. But they can also be a problem in a domestic garden, with mice sneaking into your shed to snack on your pea and bean seeds, or rats devouring your prize sweetcorn. Rodents hate pungent smells, so you could experiment with these old-fashioned tricks to deter the intruders:

- Roll pea and bean seeds in paraffin.
- Soak a few rags in tar (you could get some from a friendly roofer) and leave them lying around.
- Grow mint or soak cotton wool balls in peppermint oil and arrange strategically.
- Scatter mothballs where rats and mice lurk.

Rabbits

But Peter, who was very naughty, ran straight away
to Mr. McGregor's garden, and squeezed under the
gate! First he ate some lettuces and some French beans;
and then he ate some radishes; And then, feeling
rather sick, he went to look for some parsley.

Beatrix Potter, *The Tale of Peter Rabbit* (1902)

If you live in a rural area, rabbits can be a real pest. It is said that they dislike foxgloves and onions, so grow these near your treasured plants. They also seem to dislike lavender, rosemary, thyme, sage and lemon balm.

Keeping diseases at bay

Many of the strange spots or markings you might see on your plants are the result of fungal infections. These reduce the vigour of plants and make them look unsightly. Four of the most common fungal diseases are:

Powdery mildew – a white powder on leaves, also on stems and buds

Black spot – yellowing leaves covered with black spots; the leaves then drop; common on roses

Rust – rusty-brown spots on leaves; especially common on hollyhocks

Brown rot – creamy-white spots followed by rust-coloured rotting and shrivelling fruits; affects many stone fruit, including apples, plums, peaches and apricots.

Good garden practice

Chemical fungicides are available for some diseases, for example, black spot. But if, like the gardeners of old, you prefer to rely on good garden practices rather than chemicals, there are steps you can take at least to prevent infection taking hold.

- Remove affected leaves and burn them, or destroy them in some other way that will prevent new infections by dormant spores. **Do not add them to your compost heap!** Dormant spores can contaminate the compost so that it leads to new disease when spread on the soil.
- Lay a thick mulch to bury any infected leaves that have fallen onto the soil and to suppress overwintering spores. In spring, hard-prune the shoots of plants affected the previous year, and burn what you remove.
- Remove any fruit infected with brown rot as soon as you see it, to prevent the disease spreading from one fruit to the next. You can bury the fruit – but the hole you dig will need to be at least 30 cm (1 ft) deep – or you may be able to add to your local authority's green waste collection (check with them first).
- Encourage good air circulation and water adequately to discourage powdery mildew. This disease thrives where there is dryness at the roots and warm, damp air around the top of

the plant. Space out your plants and locate them away from walls where the air is still. Keep the centres of the plants open by pruning. Water well in dry spells, and mulch the soil to retain moisture.

- Keep your tools clean so you do not spread the disease.
- Act early to nip disease in the bud.

Milk against mildew

Who would have thought it? When cow's milk is exposed to sunlight, it's effective against powdery mildew – and it can also fertilize your tomato plants at the same time.

Mix an equal quantity of milk and water. Spray the mixture on the leaves once a week, early in the morning. Allow to dry for 30 minutes, then wipe off any excess.

Works on: courgettes (zucchini), tomatoes, grapevines

Garlic against black spot

Grow garlic and chives beneath roses, as they are said to inhibit black spot.

Rhubarb against clubroot

An old gardener's trick to curb clubroot in cabbages is to bury a few rhubarb stems in the soil where these brassicas are to grow.

Mothballs

Hang mothballs in peach trees to prevent leaf curl.

Meths

Spray cabbages and sprouts with methylated spirits against mildew – though you might want to pick off the outer leaves before you eat them.

Pea-soupers

Sulphur – the Bible's dreaded 'brimstone' – is a natural inhibitor of black spot and mildew. In the old days, before laws were passed to control air pollution and chimneys were belching out smoke from coal fires, the levels of sulphur dioxide in the air were high. London was particularly badly affected, with its infamous smogs known as 'London particulars' or 'pea-soupers' because of their greenish colour, caused by soot particles in the air.

Things got especially nasty in December 1952, when damp, freezing weather, very little wind and high pressure shrouded the city under a layer of dense, polluting fog in what became known as the Great Smog. Visibility was reduced to around 1 m (3 ft) and twelve thousand people are said to have died. But there was a small silver lining to such sulphur-rich atmospheres: back then, roses suffered from a lot less black spot, to which they are especially prone. Today, eco-safe products containing sulphur are available for the treatment of black spot and mildew.

9

Gardening Without a Garden

Some of the most delightful of all gardens are the little strips in front of roadside cottages.

Gertrude Jekyll (1843–1932), British horticulturist,
garden designer and writer

You might drool over pictures of gorgeous gardens and fantasize about one day having your own plot of land to turn into something beautiful, but what do you do if the only outdoor space you have is a high-rise balcony? What if all you have is a window ledge? It is true that getting on out there, up close and personal with the soil and all those lovely earthy smells, is hard to beat. But if your space is very limited, there are still many ingenious ways you can 'garden without a garden'.

As well as such obvious solutions as pot plants and window

boxes, being a 'no-garden gardener' requires a switch in consciousness: the ability to see growing potential in the most unlikely of places. Even in cities, nature is all around us. There are grass verges and uncultivated railway banks; there are plants that have stubbornly taken root in cracks in concrete or paving stones.

Whatever your resources as a gardener, remember one key fact: plants *want* to grow. Your job is simply to facilitate that drive for life. The rest is just the detail of how you do it. Much of the sensible, old-fashioned advice given earlier in the book also applies to the no-garden garden.

> *Every blade of grass has its angel that bends*
> *over it and whispers, 'Grow, grow'.*
>
> The Talmud, the second-century book of Jewish law and teachings

Grow in containers

This is the obvious first choice if you have no garden. Terracotta is the gold standard for planters and can look extremely good. On the down side, terracotta pots can be extremely heavy, crack easily, are porous (if unglazed) so that the soil in them dries out more quickly, and they can be eye-wateringly expensive. If you don't want to splash out on terracotta, you can – in the canny and frugal spirit of a true old gardener – choose the re-purposing option.

Almost any container can be pressed into service as a planter, as long as it can hold the required amount of compost. It will need drainage holes in the bottom; if there aren't any, make some with a drill or, for thinner forms of plastic, melt holes with a hot skewer so that the plastic doesn't split. Don't forget to lay some crocks (broken bits of pottery) or small stones over the bottom to assist drainage, then fill your container with potting compost. So get creative and see what your imagination can come up with.

Builder's buckets On sale in hardware and DIY stores and builder's merchants, these make good, sturdy planters. They are usually black, which provides a good foil for colourful plants.

Plastic waste bins or small refuse bins The advantage of plastic is that it is waterproof and so retains moisture for longer than terracotta; however, it doesn't keep plant roots as cool as terracotta.

Stacking storage boxes These rectangular containers can be fairly large, allowing you to combine more plants together for a more dramatic display.

Galvanized buckets These can bring a stylish, rustic look to your no-garden garden. Bear in mind, though, that metal heats up, so it would be wiser to place the bucket in a cooler, shadier spot. Metal eventually rusts, too, unless you seal it first.

Old paint tins Remove all traces of paint and clean thoroughly inside. The straight-sided shape suits modern garden style; or you can paint the outsides in pastel or bright colours – or even patterns – for a cheerful look.

Wicker baskets These make delightful, country-style planters, with an attractive woven texture. Line them with polythene, not forgetting drainage holes. Raise them off the ground by placing them on bricks to protect them from damp and prevent them from rotting.

Old-timer's tip

Compost in containers dries out more quickly
than open soil. To avoid the need for constant
watering, grow plants together in larger pots
or boxes, which retain moisture for longer.

Build a tower

This clever idea makes the most of minimal ground space by
stacking plant pots on top of each other. Start by filling a large
pot with compost, then place another slightly smaller pot on top;
set it in the centre, leaving a ring of uncovered compost around
the edge of the lower pot. Keep building your tower, adding a
successively smaller pot each time, until you have three more tiers
of pots. You can now plant into the rims of compost in each tier.

Think about design

Even if you're working with a very small space, you still need to
think about design and arrangement – perhaps even more so in
this case, because your eye will be more focused. Here is a list of
things to consider:

**The size, shape and style of your container in relation to
the plants it will hold.** You could go for contrasts in shape
and texture, for example, such as rustic wicker with more

architectural plants; or plain, straight-sided paint pots with frothing, colourful cottage garden flowers.

Colour: will you go for a container that subtly echoes the colour of the plants, or would you prefer strong contrasts?

Groupings: Traditional wicker baskets, for example, are not natural partners for more contemporary shapes, such as plastic waste bins.

The arrangement: Cluster your planters together in a pleasing group, with the taller ones at the back, or raise pots on shelving (see 'Garden shelving' on page 173).

Try window boxes

In the Mediterranean, people don't let a little lack of space hold them back. Just think of all those fiery red geraniums (correctly, 'pelargoniums') tumbling out of window boxes and spilling from balconies that so typify countries such as Greece, Italy and southern France.

Unless your window ledge is wide enough to fit a window box, you'll need to fix brackets to the wall to hold the container. You can fill your window box with compost and plant directly into it, or you can place smaller pots inside it.

If you have railings on your balcony, another clever idea is to hang pots from it. You can buy ready-made pots with hooks for looping over a railing, but it's not difficult to improvise your own. You'll need a pot with a rim (rather like a pudding basin). Tie a double loop of string or wire around the pot, just below the rim (which will keep the string in place), leaving four loose ends that are sufficiently long enough to tie to the top of the railings.

Head upwards

Another obvious solution for small-space gardening is to exploit your walls as growing surfaces. This can double, treble or quadruple the space you have available at ground level.

Hanging baskets
These can look very pretty, with flowers tumbling out of them. Try not to locate them so high that you can't easily reach them for watering, and remember to line them to retain moisture. Sphagnum moss is the traditional material here, but if you want to be *truly* frugal you could use a piece of an old woollen sweater, or even newspapers, although these will rot down after a while.

Back in time

The healing powers of sphagnum moss have been known for centuries because of its antiseptic and absorbent properties. Ancient Irish warriors used this moss to pack their wounds, and it was employed for the same purpose during World War I.

Garden shelving

You can fix shelves to walls or you can buy purpose-made, freestanding shelving units on which to place your plant pots. But why buy when you can do some sensible, and free, recycling?

A small wooden box, attached to the wall, will do as well.

Some spare bricks and a couple of wooden planks will provide the raw materials for the easiest shelving system of all, which can be assembled in minutes without the aid of any tools. Just place the planks on little stacks of bricks and build up the unit in that way, remembering to leave enough vertical space between the planks for your plant pots.

An old wooden ladder or redundant wooden bookcase are other options. Don't forget that wood needs sealing with paint or varnish to protect it if it is being left in the open.

Metal shelving units – the kind people might use in sheds or garages – will give a more contemporary, urban look if that if what you want. Metal needs protecting, too, so you could paint it first and keep an eye out for rust.

Green walls

Also known as living walls or vertical gardens, these are walls covered with greenery and plants. They are different from the traditional ivy-clad surface, which involves just one very large climber; these are formed from masses of individual plants which are attached to the façades of tall buildings, to clothe them in a swirling, patterned carpet of greens. The best are works of art in their own right. They are also said to help reduce air pollution and have been enthusiastically adopted by architects and designers around the world.

You won't be able to achieve anything as spectacular as, say, the Caixa Forum living wall, covered with tens of thousands of plants and towering 24 m (79 ft) high above the centre of Madrid. But you get the idea. Create your own domestic version by using one of the vertical planters now available from garden centres and other retailers. These consist of rows of planting pockets that you hang on a wall – rather like those old-fashioned shoe hangers designed to hang from the top of a door. Indeed, the latter make good, and cheaper, substitutes, though they may not be quite as durable as the purpose-made varieties.

Other ways of achieving your green wall include fixing trellis or even part of an old wooden pallet to the wall, and attaching plant pots to it.

The no-garden kitchen garden

Just because you don't own a vast acreage, there is no reason why you can't enjoy growing your own fruit and vegetables. Space restrictions will greatly limit how much you can grow, but it's still worth it – nothing beats the flavour of freshly picked, homegrown produce and they look so decorative too. It's not difficult, either: vegetables and fruits are just plants, after all, so where you grow flowers you can also cultivate edibles.

One advantage of growing such crops on, say, a balcony or in a window box is that the nearby walls act like radiators, absorbing the sun's heat during the day and releasing it at night, and thus keeping your plants cosier and warmer. If the site is exposed and windy, you might also want to put up some form of screening to give your plants extra shelter.

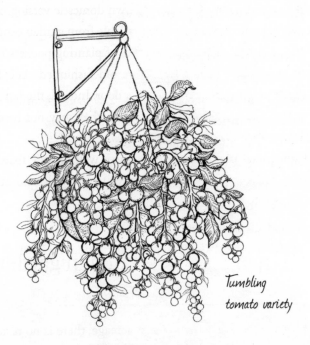

Tumbling tomato variety

Compact, trailing and climbing plants are the best candidates for your mini kitchen garden. Here are just a few ideas to get you going:

- Courgettes (zucchini) are prolific and a single plant will crop generously.
- Sweet and chilli peppers and aubergines (eggplants) are compact and manageable but need a warm spot and support with canes.

- Strawberries are great candidates for hanging baskets or a pot 'tower' (see 'Build a tower' on page 170); not only will they look pretty, but they'll do well, too.
- Tumbling varieties of tomatoes are good for window boxes or hanging baskets.
- Runner beans have pretty flowers and lush foliage and will quickly shoot up a trellis. They also make attractive screening on a balcony.
- Trailing plants such as squashes can also be grown as climbers.
- Swiss chard, with its beautiful red and yellow stems, or curly kale are ornamental enough to be mixed in with your flowers.
- Loose-leaf lettuces allow you to pick leaves singly, so they will still look decorative while they are being harvested.
- Fruit bushes, such as gooseberries or currants, are compact enough to fit on a balcony.
- Patio fruit trees, specially bred for their small size, could work on larger balconies, provided these are robust enough to take the weight.

Old-timer's tip

Grow your climber on a moveable trellis. Hang your trellis on the wall using hooks rather than screwing it in place. Then, when you need to repaint the wall, carefully lift off both trellis and plant, before hanging them back up again on the repainted surface.

Gardens in the sky

Space-poor urban dwellers often take advantage of flat roofs and turn these into roof gardens. They're a great idea and make good use of an otherwise dead space but need to be given careful thought. Is the roof strong enough to take the weight of whatever will be placed on it, or will it need strengthening? Do you need permission to convert your roof in this way? Will you need to add screening, both for privacy and safety? Always seek specialist advice before you embark on such a project.

If you fancy the idea of gazing out over a wild-flower meadow rather than a dreary expanse of asphalt, a green roof could be the answer. These come in the form of matting, ready sown with different plants such as sedums. Some are suitable for DIY, but others may require professional installation. As well as looking great, green roofs provide added insulation.

Back in time

Turf roofs have been used for centuries in traditional rural buildings. Scandinavian countries used them as far back as Viking times and beyond. The roof rafters of Irish cottages were stuffed with turf, then covered with thatch, while the neolithic burial mound of New Grange, dating back to 3200 BC, has perhaps the oldest green roof in Ireland.

Communal gardens

Opportunities to transform the world with the beauty of plants are everywhere around us, if we but take the trouble to look.

Guerrilla gardening

Have a look around your street. Are there any little corners that could be transformed into mini gardens, for example, that little patch of bare soil under a nearby tree or that unloved strip of earth bordering a row of houses? Swoop in like a horticultural crusader and get sowing and planting. You will be improving the neighbourhood for everyone, and your neighbours might like to join in the adventure, too.

Allotments

Even if you are lucky enough to have your own garden, tending an allotment is a special and totally different experience. It's something to do with the space and the sense of 'getting away from it all', as well as the supremely satisfying fact that you are growing your own food. If you don't currently have an allotment, put yourself on the waiting list for one – it will be worth the wait. (Also see 'The allotment' in The Kitchen Garden chapter.)

The urban farms of Havana

Garden bridges, city farms, organic collectives – urban dwellers the world over are finding inventive ways of greening their home towns. But there is one initiative that is particularly inspiring.

During the Cold War period in the middle of the twentieth century, Cuba relied heavily on the support of the USSR, while at the same time being subject to a trade embargo by the United States. When the USSR collapsed, Cuba's economy plummeted, agricultural production dropped dramatically and there were food shortages. Something had to be done, and it was. Out of the ashes of this disaster rose a horticultural phoenix, which remains a shining example of resourcefulness, self-reliance, survival and sustainability: the urban farms of Havana. With the encouragement and support of the Cuban government, the residents of the city got together to clear derelict spaces and transform them into small farms where they could grow their own food. Where ground was once littered with rubbish and rubble, immaculate rows of edible crops now appeared. At the foot of looming high-rise buildings or squeezed into reclaimed spaces between crumbling houses, vegetables and fruit of all kinds flourished. Where intensive agriculture on state farms was previously the norm, now ordinary people, working cooperatively, became food producers.

The citizens of Havana were still poor, however, which meant that they could not afford artificial fertilizers or chemical pesticides and herbicides, which the country had access to in the Soviet era; now they had to farm organically.

What has happened in Havana encapsulates all the values of the 'Old Wife' or 'Old Boy': waste not, want not; save money; be practical and use your common sense; don't throw anything away if you can re-use or recycle it; learn from others and share knowledge; and respect Nature.

Guide to
Botanical Names

Plant names can be a bit of a minefield at the best of times, but the common, or folk, versions can be positively bewildering. Not only can these vary from country to country, but they can also change according to region within the same country. People in rural areas often came up with names based on a plant's resemblance to something else, with some quaint and charming results. Who could resist a flower called 'granny's bonnet' or 'lady's mantle', or not be intrigued by 'love-in-a-mist', 'false goatsbeard' or 'shepherd's purse'? In the case of plants used in herbal medicine, folk names frequently related to the conditions which those plants were thought to remedy: hence, for example, 'boneset', which was believed to help mend broken bones.

To find your way through the tangled undergrowth of common plant names, the only solution is to get scientific. Back in 1758, the Swedish botanist, zoologist and physician Carl Linnaeus came up with a system for classifying the natural world which we still rely on today. Using terms from Latin and Greek, each living organism was given a two-word name: the *genus*, or larger group to which it belonged (as in *Homo*, or 'man'); and the *species*, denoting those differences that marked it out from others in the genus (as in *sapiens*, or 'wise'). Where all the species within a genus are being referred to, the abbreviation spp. is used.

Here are the botanical equivalents for the less familiar common and folk names you will find in this book:

Achillea *Achillea millefolium*
Alecost or costmary *Tanacetum balsamita*
Alfalfa *Medicago sativa*

Betony *Stachys* (or *Betonica*) *officinalis*
Boneset *Symphytum officinale* (boneset is also the common name given to *Eupatorium sordidum*, commonly found in the US)
Buckwheat *Fagopyrum esculentum*
Bugle *Ajuga* spp.
Buttercup *Ranunculus* spp.
Butterfly bush or buddleia *Buddleia* (or *Buddleja*) *davidii*

Camomile *Chamaemelum nobile*
Canterbury bell *Campanula medium*
Catmint *Nepeta* spp.
Chervil *Anthriscus cerefolium*
Clary sage *Salvia sclarea*
Comfrey (same as boneset)
Common laurel *Prunus laurocerasus*
Coneflower *Echinacea* spp.
Costmary (same as alecost)
Cowslip *Primula veris*
Crab apple *Malus* spp.
Cumin *Cuminum cyminum*

Dame's rocket *Hesperis matronalis*
Dog rose *Rosa canina*

False goatsbeard *Astilbe arendsii*
Fennel or finkle *Foeniculum vulgare*
Figwort *Scrophularia* spp.
Fleawort *Plantago* spp.
Fox and cubs *Pilosella aurantiaca*
French marigold *Tagetes patula*

Globe thistle *Echinops* spp.
Good King Henry *Blitum bonus-henricus*
Granny's bonnet *Aquilegia* spp.
Grazing rye *Secale cereale*
Ground ivy *Glechoma hederacea*
Ground oak *Teucrium chamadrys*

Horehound *Marrubium vulgare*
Hyssop *Hysoppus officinalis*

Keys of St Peter (same as cowslip)
Knitbone (same as boneset)

Lady's mantle *Alchemilla mollis*
Larkspur *Consolida* spp.
Love-in-a-mist *Nigella damascena*
Lungwort *Pulmonaria officinalis*
Meadowsweet *Filipendula ulmaria*
Mexican marigold *Tagetes minuta*

Mock orange *Philadelphus virginalis*
Mugwort *Artemisia vulgaris*
Mustard greens *Brassica juncea*

Night-scented stock *Matthiola longipetela*

Orange hawkweed (same as fox and cubs)

Plantain (same as fleawort)
Poor man's asparagus (same as Good King Henry)
Pot marigold *Calendula officinalis*

Red clover *Trifolium pratense*
Rocket *Eruca vesicaria*
Rue *Ruta graveolens*

Sage *Salvia officinalis*
Sea holly *Eryngium maritimum*
Shepherd's purse *Capsella bursa-pastoris*
Snapdragon *Antirrhinum* spp.
Soapwort *Saponaria officinalis*
Soldier's tea (same as horehound)
Solomon's seal *Polygonatum* spp.
Sweet betty (same as soapwort)
Sweet rocket (same as dame's rocket)
Sweet william *Dianthus barbatus*

Tansy *Tanacetum vulgare*
Tobacco plant *Nicotiana* spp.

Verbena *Verbena bonariensis*

Wall germander (same as ground oak)
White clover *Trifolium repens*
Willow *Salix* spp.
Winter aconite *Eranthis* spp.
Winter wolf's bane (same as winter aconite)
Wood spurge *Euphorbia amygdaloides*
Wormwood *Artemisia absinthium*

Yarrow (same as achillea)

Index

Page numbers in *italic* refer to illustrations